HOW TO RESTORE YOUR JEEP

JEEP 1941–1986

Mark Altschuler

***CarTech*®**

CarTech®

CarTech®, Inc.
6118 Main Street
North Branch, MN 55056
Phone: 651-277-1200 or 800-551-4754
Fax: 651-277-1203
www.cartechbooks.com

Edit by Wes Eisenschenk
Layout by Monica Seiberlich
ISBN 978-1-61325-452-3
Item No. SA443

Library of Congress Cataloging-in-Publication Data Available
Written, edited, and designed in the U.S.A.
Printed in China
10 9 8 7 6 5 4 3 2 1

CarTech books may be purchased at a discounted rate in bulk for resale,
events, corporate gifts, or educational purposes. Special editions may also
be created to specification. For details, contact Special Sales at 6118 Main
Street North Branch, MN 55056 or by email at sales@cartechbooks.com.

A special thanks to Dave Logan for his help and expertise with
editing and photos.

DISTRIBUTION BY:

Europe
PGUK
63 Hatton Garden
London EC1N 8LE, England
Phone: 020 7061 1980 • Fax: 020 7242 3725
www.pguk.co.uk

Australia
Renniks Publications Ltd.
3/37-39 Green Street
Banksmeadow, NSW 2109, Australia
Phone: 2 9695 7055 • Fax: 2 9695 7355
www.renniks.com

Canada
Login Canada
300 Saulteaux Crescent
Winnipeg, MB, R3J 3T2 Canada
Phone: 800 665 1148 • Fax: 800 665 0103
www.lb.ca

CONTENTS

PREFACE

I still remember the first time I was interested in Jeeps. I was about 6 or 7 years old and was watching *The Roy Rogers Show* on TV. Like all of the cowboy-crazy kids in the 1950s, I wanted a horse like Trigger—the horse that was cast as Roy Rogers's horse in the show. Every kid wanted Trigger. But as much as I wanted Trigger, I wanted *Nellybelle,* which was a 1946 CJ-2A that was driven by Roy's comical sidekick, Pat Brady. It had a

Author Mark Altschuler is pictured.

During World War II, Jeeps were shipped overseas in wooden crates. Few, if any, crated Jeeps were sold as surplus after the war.

half windshield and doors so that the stuntman could hide inside, drive it around, and chase Brady.

As far back as I can remember from the 1950s, 1960s, and into the 1970s, there were advertisements in *Popular Mechanics* and many other magazines and comic books about Jeeps in a crate. It sounded too good to be true—an Army Jeep in a crate for $49.95 to $149.95.

In the late 1960s, I had been teaming up with my friend Bob Ruben in Oyster Bay, Long Island, New York. We set up a shop and put out some great restorations. However we couldn't find enough Jeeps to work on, so I figured I'd need a different avenue to find some.

Out of curiosity and desperation, I sent in my $1 to get the information about this great offer. A few weeks later, I received a reply. They were offering to sell me everything under the sun: pens, clothing, toys, etc. but nothing about Jeeps

or any cars or trucks. If you wanted those products, you had to send in another dollar for more information. So, I did.

Again, in return, I received offerings for all kinds of goods. If you wanted information on surplus Army Jeeps in a crate, send in another $1. So, I did. However, they told me if I wanted to get government-surplus Jeeps to send a postcard to the government logistics office in Battle Creek, Michigan, and I could get on the bidding list.

After a 5-cent postcard, $3, and a month of my time, I was placed on the bidding list and began to receive bidding catalogs on everything the U.S. government was selling. Almost all the auctions were sealed bids. All that was shown was the description and some photos in the catalog. I had to familiarize myself with how the government described its goods.

To bid, you had to send in a 10-percent deposit on the date the government opened the bid. You had 10 days to pay in full, and then you had 30 days to remove the Jeep, Jeep parts, or any other items you may have purchased. If you didn't make the payment on time and remove the Jeeps in time, you were removed from the bidding list.

I would only bid on Jeeps on the East Coast within a 10-hour drive of New York. By the time you factor in the trucking fees and the missing parts, it was a gamble. In 1967, I flew to Norfolk, Virginia, to a rare walk-in auction. I bid on five M38-A1s. The Jeeps were actually in Fort

Beauvoir, Virginia. I bought them for $125 each and had to truck them home for an additional $100 each. Out of the five Jeeps purchased, I made three complete Jeeps. I ended up with many extra parts, lots of work, and not much profit.

You can still buy them through government auctions today. A friend of mine bid on a 1953 M38-A1 in Washington state. He bid more than $4,000 on the Jeep plus the shipping

fees to deliver it to Nevada. It ran but needed brakes and a few other repairs.

Reward!

Sometime in the late 1970s, I became tired of all the marketing materials telling me about the Jeeps in crates. I put an advertisement in the *New York Times*: "$1,000 reward for anyone who can produce or

show me a surplus Jeep in a crate." There was always someone who would claim to know a guy who had them, but you had to buy 10 or 50 at a time, allegedly.

They did ship Jeeps all over the world during World War II. If you find a World War II Jeep or M38 that you would like to restore, chances are good that it was shipped back to the United States from Europe or Korea. Good luck in your search.

INTRODUCTION

If you have a 1941–1945 MB or GPW, you should not have the name "JEEP" on the engine head. It should just say "Willys." The name Jeep did

not come about until the 1946 models. During World War II, the service men loved them. The government official name for them was the Gen-

eral Purpose Utility Truck, and the service men shortened it to GP. By saying it out loud, you can see how "GP" turned into "Jeep."

Civilian Jeeps evolved with each new model, and many of the parts were interchangeable. It is fairly common to find Jeeps with parts from older or newer CJ models.

There are a few other theories as to how Jeep got its name, but this theory is most widely accepted.

Before you start a restoration, you need to know about the pizza Jeep. It is not a Jeep that delivers Dominos pizza with pepperoni and extra cheese. It's a Jeep that is a "pizza of this" and a "pizza of that."

Many Jeep models are very close in style, and the parts are interchangeable. So, over the past 50 years, some strange combinations have developed. It's very common to see a 1946 Jeep with a 1952 windshield. So, if you were to order a new windshield wiper motor for your Jeep, which one would you order? Is it vacuum or electric? When you're going to restore a Jeep, it is very important that you can identify the year, make, and model.

Sometimes it is easy to fool even the experts if you're not paying attention. The following is a great example.

Pizza Jeep Project

This Jeep was restored by my neighbor about 10 years ago. He passed away, and it sat for five years. I drove by many times and finally was able to buy it. It was presented to me to be a 1941 MB. What a great find. It had no rust and was complete.

The first problem is that it has a contemporary Nevada open title. The title declares this Jeep as a 1941. The serial-number tag on the inside front of the chassis and the number on the plate on the dash match. So, what is the problem? The title indicates the Jeep is a 1941 model, which would be rare because only about 8,000 were made. However, the serial numbers on the Jeep are in

the 300,000 range, which indicates a 1944 model. It's still a good find but not correct.

The second problem is that the photos of this Jeep in the listing looked great. It was a complete, nice-looking, World War II MB. A closer look reveals a CJ-3A gas tank, no rear toolboxes on the inside wheel wells, and no glove box on the dash. There is more, but it tells me that someone took a 1940s MB that most likely had rusty floors and installed a tub from a CJ-2A or 3A. This is the worst pizza Jeep I've seen in a long time.

If you were looking for a nice "run around" Jeep, this would be okay. However, if you wanted to restore this back to the original World War II condition, you would have to replace the tub, rebuild the engine, and spend about $10,000.

A BRIEF HISTORY OF THE JEEP

In 1919, the U.S. government realized after its experience in World War I that it needed a new type of vehicle. It wanted a general-purpose all-terrain vehicle to carry two to four men and supplies over rough terrain. It had to be four-wheel drive with high ground clearance. But, the government didn't have the budget for the vehicles at the time.

However, on July 11, 1940, as World War II began, the government wrote specifications and sent out request for proposals (RFPs) to 135 American automotive companies. Bids were to be received back by July 22.

Bantam Car Company

Bantam delivered its first prototype vehicle for testing on September 27, 1940. The U.S. Army Quartermaster Corps liked it, but Bantam did not have the facilities to meet the contract demands, so it was not included in the government contracts. The company ended up building trailers for the Jeeps instead.

Willys Overland

As with most inventions, a number of prototypes were developed before the final product came about. Willys' first prototype, the Quad, was delivered for testing in November 1940 to obtain a government contract.

Willys learned from the Quad and quickly moved on to its model MA, ironing out many of the problems and including some improvements. Willys, satisfied with what it had in the MA, submitted its creation for the government contract. On July 1941 after the army requested further changes, Willys received a contract from the government to make the Willys MB general-purpose utility truck.

Ford Motor Company

Ford was reluctant to join the race for this government contract but

The American Bantam Company delivered the first prototype to the Army Quartermaster at Camp Holabird, Maryland, on September 21, 1940. The Bantam BRC was tested until it was destroyed. The Willys-Overland Quad and Ford Pygmy were delivered for testing in November 1940.

eventually did. In November 1940, it submitted its Pygmy for testing. Refinements after testing resulted in the production of the Ford GP prototypes. Ford called its production Jeep the GPW. The "G" stands for "general," the "P" stands for "Ford passenger line," and the "W" stands for "Willys design."

G503

During the war, the Willys MB and Ford GPW needed to have interchangeable parts for field repairs. Although there are minor differences between the two models, they were able to be repaired in the field using one common set of parts.

The MB and GPW were listed in the Army catalog as part number G503. At first, both Willys and Ford branded their Jeeps by stamping the Willys and Ford name on the lower left tail panel. Ford also added its script "F" on many parts, including every bolt head and on the tops of the pistons. The Army Quartermaster insisted that both companies stop branding their vehicles. These were simply Army Jeeps—not civilian cars. Ford "F" script parts are worth a premium.

From 1941–1945 Willys produced 363,000 MBs while Ford produced 280,000 GPWs.

Making a Living on Jeeps

Sarafan Auto Parts in Spring Valley, New York, was one of the largest suppliers of Jeeps and Jeep parts. I found them when I was in high school around 1964. I would buy parts for my Jeeps from them, and they were looking for old 4-cylinder Jeep engines for rebuilding. I would drive around Long Island, see an old rusted Jeep in a gas station, and try to buy it. I would get them for $200 to $300 and drag them home. I would sell the snow plow for $300 and hook my tow bar to it and deliver the rest of the Jeep to Sarafan, which would buy the carcass from me for $150 to $200.

I got to know the folks at Sarafan, and even today, almost 50 years later, I still talk to Mikael Sarafan who supplied a few photos for this book. Sarafan Auto Parts started in 1946. Mikael's father, Irving, set up the parts business when he came out of the U.S. Army after World War II. Later, Mikael and his brother, David, joined and kept it going. This com-

pany is mentioned because it went all over the world and shipped Jeeps back to the United States. It purchased Jeeps from other governments and from members of the North Atlantic Treaty Organization (NATO).

For three years in the late 1970s, Mikael bought 50 Jeeps from NATO, and when the ship was unloaded on the Jersey docks, 12 of the 50 were shipped to my shop in Milford, Pennsylvania. They were from a NATO rebuilding depot in France or Holland. They had started to be taken apart to overhaul them to keep them battle ready, but then they were deemed obsolete and sold off. They had no transmissions or steering columns. I bought old Jeeps that were rusted from farmers and used the parts to finish the NATO Jeeps. Mikael and Sarafan Auto were responsible for returning almost 1,000 Jeeps to the United States over the years.

There is much more I could write about the wartime history of Bantam, Willys, and Ford. You'll find many books about the history of military Jeeps. However, this book concentrates on restoration, so let's focus on restoring and preserving this remarkable part of history.

CHOOSING THE RIGHT JEEP

Before you buy a Jeep, answering the following questions will help you determine the depth of project you want to pursue.

- Have you looked over parts catalogs and the internet to help forecast what you're likely going to spend on the restoration or rehabilitation?
- Have you created a detailed budget and timeline?
- Do you want to do a full ground-up restoration?
- Do you want to only do a partial rehabilitation?
- Do you have a workplace for this project that you can use for months or possibly years?
- Do you have the proper tools?

Once you find a Jeep, make sure that it has a clear title before you complete the purchase. Also, verify that the vehicle identification number (VIN) matches the Jeep's body.

With each iteration throughout the Jeep's production years, subtle changes and improvements took place. If you're having difficulty identifying a candidate for restoration (or what your current Jeep is), follow the guide below.

Willys MB and Ford GPW (1941–1945)

Model	Production Total
Willys MB	361,339
Ford GPW	280,000

The Willys MB and Ford GPW are iconic and recognized around the world. They remain popular to this day. They were built quickly and were expected to have a short life span. People in the 1940s would have been amazed to learn that these vehicles are still around and are so popular. The MBs and GPWs were built from late 1941 through 1945. (Photos Courtesy Kaiser Willys)

The classic World War II Army Jeep started it all. Designed as a utility vehicle to carry two to four men over tough terrain, this model began the love that people have for the Jeep that has lasted almost 80 years. This is the ultimate toy and prized by collectors and restorers alike. It's easily recognized by its bodystyle and the axe and shovel brackets on the driver's side, and it has handles around the side and rear of the body.

Main Identifying Characteristics
- Flat front fenders
- Nine-slot stamped grille (early MBs had a slat grille)
- Rectangular two-piece (split) outward-pivoting windshield
- Split combat wheels
- Blackout lights
- Machine gun mounting plate
- Front axle–mounted pivot arm
- Spindle arms separate from the spindle housing
- Tilting headlights
- Fuel filler under driver-side seat
- No tailgate
- Hand-operated windshield wipers
- Pioneer tool (axe and shovel) indentations on the driver's side

Technical Details	
Item	**Specification**
Length	132¼ inches
Width	62 inches
Height (top up)	69.75 inches
Height (top down)	52 inches
Tire size	6.00-16 (30 psi)
Electrical	6 volt
Engine (all years)	4-134 L-head motor (463)
Transmission	3-speed T-84
Transfer case	2-speed Dana 18
Front axle	Dana 25
Rear axle	Dana 27
Wheelbase	80 inches
Other	Two-piece windshield; no tailgate
Serial number location	Passenger's side of the dash (stamped on military data plate) and upper-left frame rail between the gusset and bumper
Engine number location	Stamped on an outcropping (boss) on the right front of the engine block behind the water pump. For the GPW, look for a Ford script "F" cast into the head and block.

MB Model Features
- Tubular arched front crossmember
- Rear shock brackets are open-channel type
- Solid toolbox lid
- Early models have "Willys" stamped on lower left tail panel

GPW Model Features
- Flat U-channel front crossmember
- Rear shock brackets are two pieces welded together (closed type)
- Toolbox lid is embossed
- Early models have "Ford" stamped on lower left tail panel

Willys CJ-2A (1945–1949)	Production total	214,760

Riding on the popularity of the World War II Army Jeep, Willys released the CJ-2A. "CJ" stood for "Civilian Jeep." Many GIs who returned home after the war purchased these Jeeps for recreation. These vehicles also found a place in both farm and industry because of their versatile uses. In the four years that they were sold, many had power takeoff (PTO) units.

Some upgrades from the World War II Jeeps included a stronger transmission, a stronger rear axle, a tailgate, a side-mounted gas filler neck, vacuum windshield wipers, and many accessories.

Main Identifying Characteristics
- Seven-slot stamped grille with flush-mounted headlights and parking lights
- Rectangular two-piece (split) outward-pivoting windshield
- Top-mounted vacuum windshield wipers
- Tailgate
- Fuel filler on outside driver's side of body
- Toolbox compartment under the passenger's seat
- No glove box
- Spare tire mounted on rear passenger's side

Willys-Overland pivoted after World War II and began building vehicles for civilian buyers. So, the company took the Army Jeep and modified it for agricultural use. The CJ-2As were marketed for farm and ranch use. Many agricultural implements were built for a variety of tasks.

One advantage Willys had was that many soldiers had driven the Army Jeeps during the war, and they understood the benefits of four-wheel drive (4WD). The CJ-2A was built from late 1945 through 1949. (Photos Courtesy Kaiser Willys)

Model Features

- 1945 to mid-1946: military indentions for axe and shovel, PTO, column shift, military-style windshield adjusting arms, short back seat, front bumper gussets, recessed parking lights
- Mid-1946: three-point hydraulic system, floor-type shifter, longer back seat, "lollipop" style of windshield adjusting arms, steering bellcrank attached to front crossmember
- 1946 to early 1950: CJ-2A bodies were fitted onto 3A frames as part of the transition into CJ-3A production

Technical Details	
Item	**Specification**
Length	130⅛ inches
Width	59 inches
Height (overall)	69 inches
Tire size	6.00-16 (front 26 psi, rear 28 psi)
Electrical	6 volt
Engine (all years)	4-134 L-head motor (463)
Transmission	3-speed T-90
Transfer case	2-speed Dana 18
Front axle	Dana 25
Rear axle	Dana 41
Wheelbase	80 inches
Other	Two-piece windshield; tailgate
Serial number location	Inside engine compartment on the upper passenger-side firewall
Engine number location	Stamped on outcropping (boss) on the right front of the engine block behind the water pump

Willys CJ-3A (1949–1953)	Production total	131,843

The CJ-3A was essentially a chance for Willys-Overland to update the Jeep to address flaws that had become apparent over time. It was an evolutionary change, not a revolutionary change. The CJ-3A was built from 1949 through 1953. In 1953, Willys-Overland was sold to Kaiser Motors, and the company was renamed as the Willys Motor Company. (Photos Courtesy Kaiser Willys)

The CJ-3A has the same body-style as the CJ-2A with some upgrades, including a one-piece windshield with center vent. They were still used primarily for work, but they were becoming more popular as recreational vehicles. Many had steps on both sides of the Jeep, and a rear seat was offered as an accessory. The VIN can be found on a small metal plate on the firewall.

Main Identifying Characteristics

- One-piece windshield with rounded glass and center vent
- Bottom-mounted windshield wipers
- Spare tire on rear passenger's side
- Driver-side seat lowered and moved back for more legroom/headroom

Special Models

The 1951-and-later CJ-3As could be ordered as a Farm Jeep, which was equipped with equipment such as hydraulic lift, PTO, drawbar, propeller shaft guards, heavy-duty springs, and variable-speed belt-driven governor. A Jeep Tractor was another special model. Both were available as a CJ-3A or CJ-3B. The VIN looked like this: 451-GC1 10001. The "C" indicated it was a Farm Jeep, and a "D" indicated it was a Jeep Tractor.

Technical Details	
Item	**Specification**
Length	129¾ inches
Width	59 inches
Height (overall)	66¾ inches
Tire size	6.00-16 (front 26 psi, rear 28 psi)
Electrical	6 volt
Engine (all years)	4-134 L-head motor (463)
Transmission	3-speed T-90
Transfer case	2-speed Dana 18
Front axle	Dana 25
Rear axle	Dana 44
Wheelbase	80 inches
Other	One-piece windshield with center vent; tailgate
Serial number location	Inside engine compartment on the upper-right-side firewall
Engine number location	Stamped on an outcropping (boss) on the right front of the engine block behind the water pump

Willys M38 (1950–1952)

Production total 45,473

Most of the MBs and GPWs were left overseas when the war ended, and many of them were destroyed. So, when the Korean War began, the U.S. military needed more Jeeps. The new design included a 24-volt electrical system and was called the M38 (or MC). It was built from 1950 through 1952. This model was basically a militarized CJ-3A. (Photos Courtesy Kaiser Willys)

The VIN is located on the passenger-side dash on a stamped-brass data plate.

The M-38 was produced from 1950 to 1952, and 45,473 were manufactured. The military took what it learned from World War II and beefed up the Jeep. The United States was part of NATO at this time, so it used the standard NATO 24-volt waterproof electrical system. Based on the CJ-3A, this Jeep was improved with a larger fuel filler neck for quicker refueling along with other improvements.

Main Identifying Characteristics
- One-piece windshield
- Vacuum-style bottom-mounted windshield wipers
- Flat fenders
- Pioneer tool (axe and shovel) indentations on the passenger's side
- Waterproof 24-volt electrical system
- Protruding headlights with guard wires
- Blackout lights
- Larger 4½-inch fuel filler on driver's side

- Hinged grille
- Tailgate
- Rear-mounted spare tire
- Glove boxes
- Snorkel system
- Rear bumperettes

Model Features

The M38 is based on the civilian CJ-3A with improved characteristics, such as a waterproof electrical system, blackout lights, and a larger fuel filler for quicker refueling during combat.

Technical Details	
Item	**Specification**
Length	133 inches
Width	62 inches
Height (top up)	74 inches
Tire size	7.00-16
Fuel capacity	13 gallons
Electrical	24 volt
Engine (all years)	4-134 L-head motor (463)
Transmission	3-speed T-90
Transfer case	2-speed Dana 18
Front axle	Dana 25
Rear axle	Dana 44
Wheelbase	80 inches
Other	One-piece windshield; 24-volt waterproof electrical
Serial number location	Inside the engine compartment on the upper-right-side firewall and on a tag behind the bumper riveted to the left frame rail
Engine number location	Stamped on an outcropping (boss) on the right front of the engine block behind the water pump

Willys CJ-3B (1953–1964)

Production total | 196,000

This is called the transition Jeep. It was the same as a CJ-3A but with a more powerful overhead valve (OHV) engine that was later used in the CJ-5s. To make the engine fit, they simply raised the cowl, grille, hood, and dashboard about 4 to 5 inches to accommodate the taller engine. After 1954, the CJ-3B dashboard used the same CJ-5 speedometer gauge and accessories. It also used the shorter one-piece windshield similar to the CJ-5. There was no glove box. The VIN is located on the firewall.

Main Identifying Characteristics

- High hood
- New-style grille to accommodate the OHV Willys F-head Hurricane motor
- Flat front fenders
- One-piece windshield with shorter front frame panel
- Spare tire mounted on rear passenger's side
- No glove box
- Parking brake lever mounted in the middle of the dash.

Model Features

In 1957, the 12-volt electrical system was introduced.

The CJ-3B was another example of an evolutionary model change. Willys wanted to introduce its overhead valve (OHV) F-head engine, so it basically took a CJ-3A design and raised the grille, hood, and cowl to accommodate the taller engine. It was built from 1953 through 1964. In 1964, the Willys Motor Company was renamed as the Kaiser-Jeep Corporation. (Photos Courtesy Kaiser Willys)

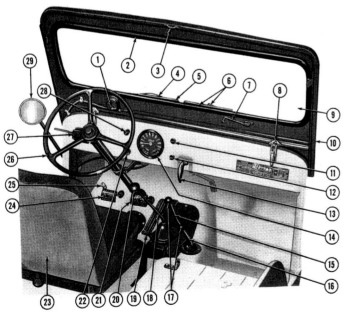

FIG. 77—FRONT COMPARTMENT (CJ-3B)

Technical Details	
Item	**Specification**
Length	129 $^{29}/_{32}$ inches
Width	59 inches
Height (maximum)	66¼ inches
Tire size	6.00-16
Electrical	6 volt (12 volt after 1957)
Engine (all years)	4-134 F-head motor
Transmission	3-speed T-90
Transfer case	2-speed Dana 18
Front axle	Dana 25
Rear axle	Dana 44
Wheelbase	80 inches
Other	One-piece windshield
Serial number location	Inside engine compartment on the upper-right-side firewall
Engine number location	Stamped on an outcropping (boss) on the right front of the engine block behind the water pump

Willys M38-A1 (1952–1971)	Production total	101,488 (80,290 domestic and 21,198 foreign sales)

The next evolution of the military Jeep beyond the M38 is the M38-A1. This is the military edition of what later became the CJ-5. It had the same 24-volt waterproof electrical system as the M38. The M38A1 is larger and more modern design than the M38.

The VIN is located on the passenger's side of the dash on a stamped-brass data plate.

Main Identifying Characteristics

- Rounded front fenders
- Two-piece windshield
- Top-mounted wipers
- 24-volt waterproof electrical system
- Blackout lights
- Rear-mounted spare tire
- Recessed headlights
- Large 4½-inch fuel filler on driver's side
- Large circular indention on the passenger-side body panel for electrical hookup
- Machine gun mount
- Dual battery box on cowl under windshield
- Glove box located on driver-side dashboard
- No tailgate
- Cut-out on hood for a snorkel
- Parking brake handle between front seats.

With a larger body, rounded fenders, and an OHV engine, the new military model was called the M38-A1 (or MD). It was built from 1953 through 1957. Exports were sold through 1971. This model was the precursor to the civilian CJ-5. (Photos Courtesy Kaiser Willys)

Model Features

The Jeep M38-A1 from 1952 to 1960 had a Dana 25 front axle, and those from 1961 to 1971 had a Dana 27 front axle.

From 1952 to 1953, the battery box was secured with eight thumb screws, and the model had a hinged front grille, freestanding radiator mount, and front fender seam at a 45-degree angle.

For 1953, the battery box was secured with a latch strap, the grille was fixed with one large bolt at the bottom center, two support rods were added to the radiator and firewall, and the angled fender seam was replaced with a vertical seam.

Technical Details	
Item	**Specification**
Length	138 $^5/_8$ inches
Width	60 $^7/_8$ inches
Height	73 $^3/_4$ inches
Tire size	7.00-16
Electrical	24 volt
Engine (all years)	4-134 F-head motor (475)
Transmission	3-speed T-90
Transfer case	2-speed Dana 18
Front axle	Dana 25, Dana 27
Rear axle	Dana 44
Wheelbase	81 inches
Other	Two-piece windshield
Serial number location	Passenger's side of dash (stamped on brass data plate)
Engine number location	Stamped on an outcropping (boss) on the right front of the engine block behind the water pump

Willys/Jeep CJ-5 and CJ-6 (1955–1983)

Model	Production Total
CJ-5	606,303
CJ-6	50,172

The only difference between the CJ-5 and CJ-6 is the length of the wheelbase. The CJ-5 has an 81-inch wheelbase and the CJ-6 has a 101-inch wheelbase. The final step in the evolution of the basic Jeep from 1941 to 1983 was the CJ-5. This model was very popular. If you can find one in decent condition and is basically original, this model is a great candidate for restoration and is highly sought after.

Main Identifying Characteristics

- Rounded front fenders
- Seven-slot stamped grille
- Heavier rounded hood
- One-piece windshield with top-mounted wipers
- Windshield hinged to top of cowl
- Spare tire mounted on rear passenger's side
- Defroster
- Glove box
- Thicker/larger seats
- 3-inch fuel filler cap
- Tailgate

Special Models

The 1961–1964 Tuxedo Park Mark III and Park Mark IV were introduced as the CJ-5A and CJ-6A models. Chrome bumpers, hood badges, column shift, an "Indian-Ceramic" colored steering wheel, and other accessories distinguish this trim package.

Technical Details	
Item	**Specification**
Length	135 $^9/_{16}$ inches (CJ-5); 155½ inches (CJ-6)
Width	71³/₄ inches (CJ-5); 72 inches (CJ-6)
Height	67 inches (CJ-5); 68 inches (CJ-6)
Tire size	6.00-16
Electrical	6 volt, 12 volt
Engine	4-134 F-head motor from 1955 to 1971; Buick 225 Dauntless V-6 motor from 1966 to 1971; Perkins 4-cylinder diesel, GM Iron Duke 4-cylinder, AMC 232 and 258 6-cylinder and 304 V-8 engines from 1972 to 1983
Transmission	3-speed T-90
Transfer case	2-speed Dana 18
Front axle	Dana 25, Dana 27, Dana 30
Rear axle	Dana 44, Model 20
Wheelbase	81 inches (CJ-5: 1955–1971); 83½ inches (CJ-5: 1972–1983); 101 inches (CJ-6)
Other	One-piece windshield; rounded front fenders and hood
Serial number location	VIN located on small metal plate screwed to firewall
Engine number location	Stamped on an outcropping (boss) on the right front of the engine block behind the water pump

In 1955, the new CJ-5 model debuted. It was essentially an M38-A1 with a 12-volt electrical system and civilian paint colors. It retained the dual battery compartment in the cowl and the cut-out hood panel for a snorkel. Jeep didn't feel the need to retool the body for civilian use and simply kept the military features. The CJ-5 was built from 1955 through 1983.

The CJ-6 was an extended version of the CJ-5 with a 20-inch-longer wheelbase. It was built from 1955 through 1981. In 1970, the Kaiser-Jeep Corporation was sold to AMC and became the AMC Jeep Corporation. (Photos Courtesy Kaiser Willys)

Jeep CJ-7 and CJ-8 (1976–1986)

Model	Production Total
CJ-7	379,299
CJ-8	27,792

The primary difference between the CJ-7 and CJ-8 is the length of the wheelbase. The CJ-7 has a 93.3-inch wheelbase, and the CJ-8 has a 103-inch wheelbase. doors were available for both.

Main Identifying Characteristics

The CJ-7 has a U-shaped door opening for easier access. The longer CJ-8 has a cargo bed but is all one body. It had an optional half-cab top. Hard tops and hard

Special Models

Special models were the Renegade, Laredo, Golden Eagle, Golden Hawk, Limited, Jamboree, CJ-8 Scrambler, SR/Renegade, and SL/Laredo.

Technical Details	
Item	**Specification**
Length	148 inches (CJ-7); 177.2 inches (CJ-8)
Width	68.5 inches
Height	67.7 inches
Tire size	P235/75 R15
Electrical	12 volt
Engine	GM and AMC 4-cylinder; AMC 232 and 258 6-cylinder and 304 V-8; 4-cylinder diesel (export)
Transmission	Manual T-4, T-5, T-18, T-150, T-176, SR-4; Automatic TH400, TF-904, TF-999
Transfer case	Dana 20, Dana 300, BorgWarner 1339
Front axle	Dana 30
Rear axle	AMC Model 20, Dana 44
Wheelbase	93.3 inches (CJ-7); 103 inches (CJ-8)
Other	CJ-7 only: Quadra-Trac full-time 4WD, Trac-Lok limited-slip differential
Serial number location(s)	VIN located on small metal plate screwed to firewall
Engine number location	Varies among each engine type

LEFT AND FACING PAGE: In 1976, the CJ-7 model was introduced. The longer-wheelbase CJ-8 began production in 1981. The CJ-7 and CJ-8 were the final CJ models sold in the United States. The longer wheelbases and bigger door openings made them easier to live with. More powerful engines were introduced by AMC. Other modern features included an optional automatic transmission and air conditioning. Full metal doors and fiberglass hard tops were also well received. The Jeep CJ era ended in 1986. Chrysler bought AMC in 1987, and the Jeep Wrangler was born. (Photo Courtesy Dawn Abbott/ Inspired Passion Photography)

TOOLS AND EQUIPMENT

Tools and equipment are an integral part of restoring your Jeep. Obviously, it would be great to have a shop with every tool and every piece of equipment you need, but that may not be practical. Many of us have to work with what we have available.

I learned a long time ago that you cannot beat a professional at his own trade. That is why I send engines to a machine shop because it would be impossible for me to pay for the tooling required to rebuild an engine properly. This also applies to the sandblaster and paint booth. We would like to think we could restore our Jeeps by ourselves, but professionals are the best avenue for some of the more expensive tasks in a restoration.

Toolbox

There is an old expression that says, "It's a poor mechanic who blames his tools." On the other hand, the proper tools can get the job done better and faster. You need a toolbox with all the basic sockets, wrenches, screwdrivers, pliers, etc. The wrenches and sockets you will need for most 1941–1986 military and civilian Jeeps are standard SAE sizes. An occasional

An engine stand has adjustable arms that allow you to bolt it to the engine via the bell-housing bolt holes. This allows 360 degrees of rotation. Keep in mind that the engine can be top heavy and can swing out of control if you don't have someone to help you. A chain hoist is needed to move the engine up or down from the engine stand, and you cannot attach the flywheel or clutch while the engine is attached to the stand.

An engine crane, or better known by its nickname the "cherry picker," can be rented from most equipment rental stores or borrowed from another hobbyist. Collapsible legs make it easy to move, and the crane can lift 2,000 to 6,000 pounds on average. Jeep engines weigh approximately 400 pounds, so the cherry picker makes it look easy. The hydraulic jack can lift the engine up and out with just a few pumps. Go slow and be careful.

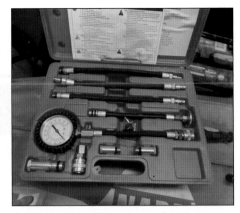

Compression testers are simple to use. Just remove a spark plug and crank over the engine. The gauge will tell you the compression in each cylinder.

If compression readings are inconsistent from cylinder to cylinder (for example, the gauge reads 50 pounds in one cylinder, 85 in another, and 60 in another), rebuild the engine.

Freezer sandwich bags are convenient for storing nuts, bolts, and small parts when disassembling the engine. Purchase bags with a labeling panel on them and use a felt-tip pen to label the contents of the bag. Many bolts and nuts look alike, and it's common for most Jeeps to have both fine and coarse threads.

Ratchet straps come in handy when transporting an engine to and from the machine shop. This prevents the engine from falling over and breaking the manifold or other bolt-on components.

metric or Torx bolt may sneak in on the newer CJ models.

A good compressor and air tools will save you time, but don't get carried away with air tools. Bolts on many older vehicles have a tendency to break off under the torque of air tools. Most smaller bolts should be removed and replaced by hand. Specialty tools such as brake tools and electrical testing tools make your job easier. Always soak your bolts with a penetrating oil, such as PB B'laster,

a few days prior to removing rusty bolts. A strong, steady pull with a breaker bar and socket may allow the bolt or nut to come off intact.

When I was a young man just starting out, I met some very successful people. One of them told me something I never forgot: "Never be afraid to go into debt for a great piece of equipment because it will always pay you back."

Gantry cranes are ideal for lifting engines and removing Jeep bodies. They come in many forms. The Gantry crane shown in the photo is telescopic for adjustable height. The dolly and chain hoist cost extra.

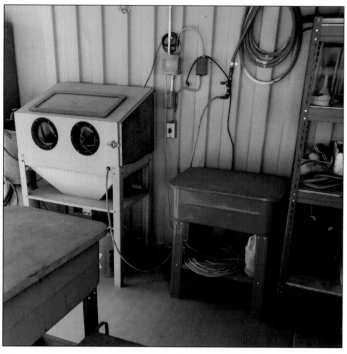

You can buy a small sandblasting machine to remove paint and rust from your small Jeep parts before you paint them. Also in this photo is a safety wash machine. Inside there is a tray and a pump to spray parts cleaner onto your parts. This makes it easier to clean off the grease and dirt.

A place is needed to organize your parts during any restoration. Label wooden boxes or plastic bins so you know where to retrieve the parts when you need them.

This engine cradle was made from a simple wooden pallet that can be moved with a pallet jack. We took a 4x4 and screwed it to the pallet. On top of the 4x4, you have to put 1x4 or 2x4 blocks to give it the height you need to attach your accessories. With the engine supported, you can attach the flywheel and clutch components.

I own many tools and equipment that I have had for 20 to 30 years, although I have had to replace some tools after the wear and tear they receive.

Favorite Tools

I have met many mechanics over the years who all say the same thing. They can have a box full of 100 screwdrivers, but they will always look for the same, trusted tool. It might not be the most expensive screwdriver, but it is their favorite.

Whether you're restoring a Jeep as a hobby or you work as a mechanic for a living, buying tools is a lifelong quest. However, what you will need to restore a Jeep is not an elaborate or expensive group of tools. Remember that most of these Jeeps were designed from 1930s technology.

This is a wooden engine stand without the engine.

The dream of all auto repair or restoration shops is to have a car lift. Lifts make life a lot easier, and using them is safer than working on the ground. Lifts typically have an electric hydraulic mechanism. Some can be purchased to run on 110-volt electricity for home use.

If you have the room and can afford a car lift, buy one. There are a variety of car-lift types available. If you don't have the room or budget, use floor jacks and jack stands. They will get the job done.

This table is very practical when removing or assembling engine accessories. The engine is at a more comfortable height. Notice the assortment of wooden blocks on the bottom shelf. They are very helpful in supporting a 4-, 6-, or 8-cylinder engine. By using this table, you can assemble the flywheel and clutch, making the engine ready for installation when you have attached all of the accessories. I like to keep the overhead chain hoist attached at all times as a safety precaution because the engine makes this table top-heavy.

In the foreground is a very handy tool cart. On it, you can lay out your tools and whatever parts you are working with. Better than a stationary table, this cart rolls around with your tools and parts, providing easy access.

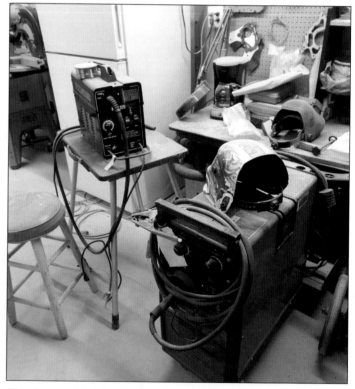

This is a welder that runs on 220 volts of electricity. The welder on the table is a small version, commonly known as a "buzz box," which runs on 110 volts of current. They are excellent for welding Jeep bodies and doing some chassis repairs. They are wire fed and easy to use. Make sure you wear proper eye protection, such as the helmet sitting on top of the larger welder.

An air compressor is one of the most important pieces of equipment in any workshop. It can be used for running air tools and blowing dust and dirt off work surfaces. You do not need a very large or elaborate air compressor to run a basic air wrench, but be sure that it produces enough airflow to run the tool. Airflow is measured in cubic feet per minute (cfm). On the floor are two portable air tanks. They can come in handy for small jobs or if your air hose will not reach your work area.

Rather than using a portable toolbox, use one of these larger roll-away tool chests. Keeping your tools clean and organized is very important, especially if you spend 15 minutes looking for a wrench to find out that it's in your back pocket.

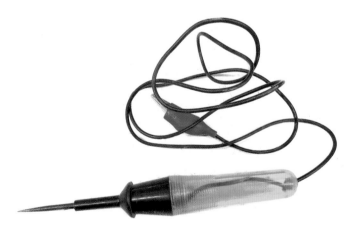

An electrical test light is helpful when working on the electrical system or repairing gauges in the dashboard. The test light indicates if electricity is flowing in the circuit. It's important to have one or sometimes two of these in case one shorts out.

A great time saver and the only way you're going to find out if you are running low on supplies is to keep them organized. There is nothing worse than wiring a dashboard only to find out you're two wiring connectors short of completing the job.

You can do a decent job restoring a Jeep with basic hand tools.

I will provide more details regarding tools and equipment as you get into each chapter. You'll see there are some specialty tools for each phase of your Jeep restoration. Auto body, brakes, and electrical are just a few components that require special tools.

Specialty Tools

There are some specialty tools that you will use just once in your restoration. Usually that tool will be

A bench grinder is a very common piece of equipment found in most workshops. One side has a wire brush, which is great for cleaning parts. The other side is a grinding wheel, which comes in handy for repairing parts and shaping your screwdrivers and chisels. Keep safety first, and always wear protective glasses.

A vise is a must for every shop. It is ideal for holding parts while assembling, repairing, and drilling holes. Many vises like this one have a flat surface that doubles as an anvil.

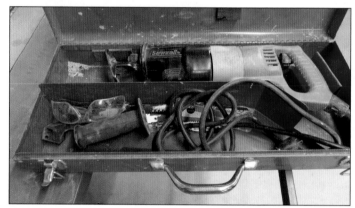

This tool is a reciprocating saw. It comes in handy when cutting off exhaust pipes or rusty bolts in hard-to-reach places. I have used it many times to cut rusty bolts out of spring hangers.

Nut drivers come in most basic sizes and save you hours of time. One use is to remove or install the nuts on the back of the gauges in the dashboard.

Socket sets come in three basic sizes: 1/4-, 3/8-, and 1/2-inch drive. Depending on what part of the Jeep you are working on, you will probably need all three sizes. Open-end and box-end wrenches come in sets of about 8 to 10 tools in the most common sizes.

residing in the bottom of your toolbox from lack of use. For example, a clutch pilot shaft alignment tool is used to install a new clutch and goes through the clutch disc and pressure plate into the pilot bushing or bearing and then into the flywheel.

Many tool manufacturers sell clutch alignment tools with various sizes of adaptors so that the tool can be used for many different types of vehicles. However, there is no need to buy this expensive tool. Many clutch manufacturers supply this tool. I personally don't rely on these plastic tools. I have an extra transmission input shaft from the front of a transmission, which I use to line up the clutch parts.

Other than the clutch alignment tool, most of the other tools you will use in restoring your Jeep are common tools that you use for other projects around your house and on other vehicles. So, don't be afraid to spend the money to buy decent tools for this project. In most cases, you will use them again and again through the years.

Keep Clean Rags Handy

A good work habit to get into is to have a good supply of clean rags along with your tools. By wiping and cleaning grease and dirt off your tools you will not become a member of the broken-knuckle club.

Besides wrenches and screwdrivers, the other shop tools you should have are brooms, rags, and a fully charged fire extinguisher. These tools will keep your shop clean and safe. They are just as important as a set of wrenches.

Always keep receptacles (buckets, pans, saucers) to catch oil or antifreeze when working with those

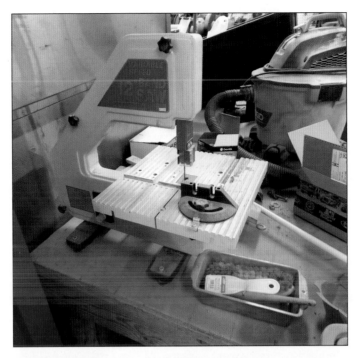

You may not believe this, but I use this band saw often to cut heater hoses and radiator hoses. You can do that with a hack saw, but I find it very useful to use this band saw because it makes straighter cuts.

There is nothing worse than to start a job or the next section in the process of your Jeep restoration only to find out you don't have the right nuts and bolts to complete the job. Keeping a variety of nuts, bolts, and supplies is essential. Keep in mind that there are two thread styles on most Jeeps: course thread and fine thread. Identify the thread pitch before trying to replace the nut or bolt. Using a thread checker indicates the size that is needed.

Items to Keep on Hand

- Engine stand
- Cherry picker
- Compression test kit
- Fire extinguisher
- Sandwich freezer bags
- Tie-down ratchet straps
- Gantry crane
- Sandblasting machine
- Shelving and storage
- Car lift and jack stands
- Tool cart
- Welding equipment and safety gear
- Air compressor, hoses, and tools
- Rolling toolbox
- Electrical current tester (6 and 12 volt)
- Electrical connector assortment box
- Bench grinder and wire wheel
- Bench vise
- Angle grinder (with grinding and cutting discs)
- Reciprocating saw
- Nut drivers
- Open-end and box-end wrenches
- Allen and Torx wrenches
- Ratchets, breaker bars, and socket sets
- Screwdrivers
- Thread-chaser kit, thread checker, and tap-and-die set
- Locking pliers and various regular pliers
- Band saw
- Nuts, bolts, and fasteners
- Hydraulic press
- Oxygen, acetylene torch set ■

parts of your Jeep. Kitty litter or other absorbents can be used for spills.

With all of the above tools and equipment, your Jeep restoration will have a head start in ensuring that you won't be wasting time running to the hardware store three times a day. Time is money, and the less time you spend chasing tools, the sooner the restoration on your Jeep will be completed.

If you don't have a cutting torch, you can buy an inexpensive kit consisting of gas pressure gauges, hoses, and a torch. Some people own their own tanks, but in most cases, you have to lease them and purchase the gas. Acetylene is a gas that will heat the metal red hot. The oxygen is pressurized to 3,600 psi. When you heat the metal red hot, you squeeze the lever that allows the oxygen to blow away the molten steel, hence the name "blowtorch." Make sure to strap the tanks to a heavy-duty cart because they can be dangerous if they fall over. Read the instructions that come with the gauges so that you do not damage their diaphragms. Try to purchase a combination torch with interchangeable ends that can cut and braze. Torches come in handy for many tasks. However, they have almost become obsolete in restoration having been replaced by angle grinders and plasma cutters.

A hydraulic press is an essential tool for pressing bearings and seals into place. It is commonly used to install a new throw-out bearing onto the throw-out bearing collar. It provides more control than using a hammer. These presses are not expensive, but they save lots of time and cause less damage.

BODY AND INTERIOR DISASSEMBLY

The disassembly process is like peeling an onion. Take the top layers off and work down until the whole Jeep has been taken apart. Having a safe place to store everything can be a challenge. Jeeps take up a lot more space when they have been taken apart. Having sturdy shelving with several shelves can be a big help and allows the storage of parts on multiple levels. Clear plastic bins can be a big help too.

Organization

When taking a Jeep apart, it's best to place the nuts, bolts, and small parts in sandwich-size freezer bags and label them. The Jeep goes back together much easier when you can find the correct parts. Taking digital photos is a good idea too.

Once the Jeep has been disassembled, wash, clean, and paint the various brackets, nuts, and bolts. This makes reassembly faster and easier.

Top and Seats

Removing the soft top (if present) is usually the first step to peeling the onion. This allows access to the interior. The top should be kept—even if its only use is as a template for the replacement top.

If you have a hard top and/or hard doors from a 1976–1986 Jeep CJ-7 or CJ-8, remove them and set them aside for cleaning and new weatherstripping. If the window regulators or door locks need repair, this can be done when convenient. It's not uncommon for the door shell to crack under the vent wing on the driver's door. This is a good time to repair that, if necessary. You should also check the door hinge bushings. If they are worn, the door may sag and not close properly. Brass replacements are available.

Next, remove the seats. Soaking the bolts with a penetrant spray

This 1953 CJ-3A is a typical example of a Jeep that has had several body modifications. A hole was cut in the dashboard for a radio, and instead of the taillights being mounted on the outside of the body, a prior owner cut large, round holes for recessed lights.

overnight may help loosen them and prevent the bolts from snapping off.

Later CJ models may have carpet. If so, it should be easy to remove it now.

Gas Tank Removal

The next step on 1941–1971 military and civilian Jeeps with the gas tank under the driver's seat is to remove all of the gas from the gas tank. Siphon it out through the filler neck or remove the drain plug on the bottom of the tank. If there is no drain plug, simply disconnect the fuel line and use a catch pan to collect the old gas.

When removing or replacing the gas tank, there are one or two hold-down straps. The military models have larger gas tanks with a 15-gallon capacity. These gas tanks sit in a well that extends below the Jeep's floor. This is a common place for dirt and moisture to collect and cause rust. It's a good idea to remove the gas tank while doing any restoration. This allows you to inspect the floor and the bottom of the gas tank.

Early civilian Jeeps have a 10-gallon gas tank that mounts to the flat floor under the seat.

After the gas has been drained, disconnect the electrical wire to the fuel sending unit on top of the tank. Then, the only things left to remove are the hold-down straps. Once they are removed, the gas tank lifts out easily.

As you remove the body bolts, determine if they are reusable. If so, place them in sandwich bags or plastic bins to keep them organized. Labeling them saves you a headache later.

This 1951 CJ-3A gas tank has one hold-down strap running from side to side. In the 1955–1971 CJ-3B, CJ-5, and CJ-6, the single hold-down strap runs from front to back. No matter which direction the strap runs, do not overtighten them to a point where you start to crush the gas tank.

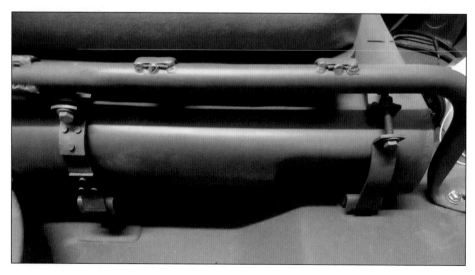

This 1944 MB shows a typical World War II military gas tank with double-strap hold downs. Notice there is plenty of extra thread in the bolts attaching the clamps. Make sure the gas tank is snug, but do not overtighten the straps.

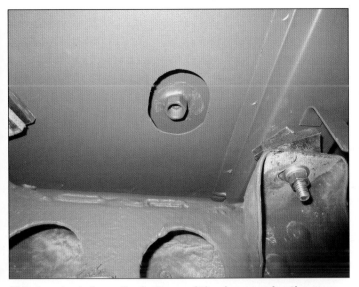

This is a look from the bottom of the Jeep under the gas tank. The fitting on the gas tank is centered in the hole on the Jeep body. You need to align the gas tank so that the fitting is in the center of the hole. This allows you to install the fuel line. If your gas tank filler neck is centered in the hole on the body and your hold-down straps are aligned properly but the hole in the floor is off-center, simply widen the hole in the floor with a file, drill, or angle grinder.

This 1968 CJ-5 shows a typical under-seat gas tank installation. The single hold-down strap runs from front to back, and the gas tank filler neck is centered in the hole on the body. Also, the gas tank filler neck is centered within the rubber grommet. The electric sending unit has the wire to the gas gauge attached with an armored cover, which runs underneath the dashboard. Take note that the screw in the sending unit is not centered. When installing this sending unit, rotate it so that all of the holes line up. It can only be installed one way.

Later-model CJs with the gas tank under the rear floor can be removed from underneath at any point in the disassembly. Siphon the gas out through the filler neck and remove the hose clamps and filler vent hoses.

Next, unbolt the gas tank skid plate. Then, support the empty gas tank with a floor jack and unbolt the hold-down strap(s). This is done by unthreading the nut(s) at the end of the hold-down strap(s). The sending unit wiring and fuel line can be disconnected as the tank is lowered. Check for other hoses on the top of the tank.

This is a good time to use a flashlight and look around inside the gas tank for rust or sludge. Gas tank cleaner is available, or you can try denatured alcohol to remove varnish. Based on your findings, you may need to clean the gas tank or replace it.

Once the gas tank is clean and empty, reinstall the gas cap on the filler neck and reinstall the drain plug or tape over the fuel-line fitting underneath to keep the tank clean until it is reinstalled.

Spare Tire, Tailgate, and Roll Bar

Another layer of the onion, the spare tire, can be removed and set aside, as can the spare tire carrier.

Early civilian model tailgates lift off their hinges once the chains are unhooked. Newer CJ models need to have the hinges unbolted and the support cables removed.

If your Jeep came with a roll bar, you should now be able to access it easily. Soak the bolts with a penetrant spray from underneath. The mounting bolts usually have a Torx-style head. You'll want a high-quality Torx

bit and a long breaker bar to apply torque to the bolts. Once the bit is firmly in the Torx head, apply a strong steady pull on the breaker bar and wait. Hopefully, the bolt will break free and not break off. Stripping the Torx head is possible too, especially if the bolt threads are rusty. If that happens, drill out the broken bolt and use an equivalent-size Grade-8 bolt and Nyloc nut when reinstalling the roll bar later. These specialty bolts and nuts keep the roll bar secure.

Windshield and Miscellaneous

You're in luck if you have an M38-A1 or 1955–1975 CJ-5 or CJ-6. Just tip the windshield forward and lift. The windshield frame should come right off the hinge. You can leave the hinge on the body and paint it with the rest of the body tub.

Now that you know a technique to remove Torx-head bolts, it's time to tackle the windshield hinges. The 1976–1986 Jeep CJ hinges use Torx-head bolts. Tapping the bolt heads with an impact driver and Torx bit works best. Otherwise, you can use a long breaker bar to apply torque. Slow, steady pressure helps. Too much torque will strip the bolt head or snap off the bolt.

It's also time to remove the taillights and gas filler neck grommet or plastic bezel, and the license-plate bracket can be taken off. Then, remove the front and rear bumpers and the black plastic frame cover on the front of 1976–1986 CJ models. This cover hides the gap between the grille and front bumper.

Footman Loops

Did you ever wonder what those little loops were that were screwed all over the Jeep bodies? They're called footman loops. You can use the webbed straps to tie down the windshield or canvas top. They also tie down folded canvas tops in the back of military Jeeps. They were originally attached with small slotted screws (not the Phillips-head screws used today). If you want to do a more authentic restoration, try to find single-slotted machine screws.

On the CJ-5, CJ-6, CJ-7, and CJ-8 models, there are two large windshield rests on the hood to support the windshield and a footman loop to strap it down. Parts suppliers sell footman loops in sets.

Early Jeeps have hood blocks and later CJ models have metal U-shaped rests to support the windshield when it is laid down. All CJ models have a webbed strap to prevent the windshield from bouncing. Removing all

the footman loops and hood catches is recommended if painting is a part of your restoration.

Side Mirrors

Over the years, these have not changed much. The 1941–1964 flat-fender Jeeps had small, round side mirrors, but the CJ-5, CJ-6, CJ-7, and CJ-8 models changed to larger square or rectangle side mirrors beginning in 1955. Jeeps had a driver-side mirror as a standard feature, but passenger-side mirrors were optional.

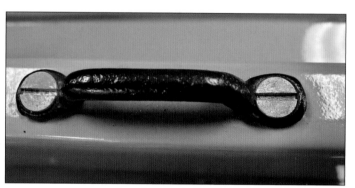

Footman loops have been around for as long as the Jeep has. During World War II, they were attached around the back of the body so that the straps from the top could be tied down. They were also used on the side quarter panels for tying down cargo. On the earlier flat-fender Jeeps, they were attached with a slotted machined screw and nut as seen here. Later, they were attached with Phillips-head screws. If the screws are rusted and hard to remove, simply use a small drill to drill them out and replace them with new screws.

On many early flat-fender Jeeps (both military and CJs), there is a double hood catch on either side. Driving over rough terrain caused the hood to flex and bounce near the back of the hood. This caused metal fatigue near the hood hinge where the hood attached to the body.

The traditional CJ-5, CJ-6, CJ-7, and CJ-8 have a small square mirror with a bracket that attached to the lower windshield hinge. The four bolts holding this bracket to the hinge go into threaded holes in the hinge. If the threads are gone, simply ream out the hole with a drill. Reach under the dashboard and put a nut, washer, and bolt into the same hole. Early CJ-5s came with a driver-side mirror, but the passenger-side mirror was optional. You will find many Jeeps with a variety of custom mirrors attached in many ways.

From 1941 to 1952, the flat-fender Jeeps came with only one style of mirror. It had a bracket bolted to the side of the body with an adjustable telescopic round rearview mirror. It's not very safe in today's world, but it was adequate back then.

Front Clip

Jeeps bodies are pretty simple, so there isn't much of a problem removing the front clip (hood, fenders, and grille). Military designers made it easy for a service man to remove the front clip for easier access to the engine to do routine maintenance or emergency repairs. The M38 and M38-A1 military models even have a hinged grille that tips forward.

Hood

The hood comes off on all flat-fender models by removing the five bolts from the hinge that connects the hood to the firewall. On the 1955–1975 CJ-5 and M38-A1 models, there is a slip hinge. Just open the hood and shift it sideways, and it comes off. On later CJ models, unbolt the two hinges on top of the hood that connect to the cowl.

On all flat-fender Jeeps from 1941 to 1953, the hood is attached to the main body by a piano-style hinge. There are five 1/4-20 bolts with washers holding on the hinge. There is also a woven metal ground strap connecting one of the hinge bolts to the body.

It is recommended that two people lift the hood and walk forward. Place a few evenly sized boards on the floor, set the corners of the hood on them, and lean the hood against the wall. Secure the hood to the wall (if possible) with a ratchet strap or bungee cords.

Hood Catches

Hood catches have been used since the very first Jeeps in 1941. One hood catch is mounted to each side of the hood to prevent the hood from fluttering. Jeeps catch a lot of wind through the grille, and that pushes up on the underside of the hood. The spring-loaded hood catches hold the hood firmly closed.

Remove the hood catch brackets on the sides of the hood. World War II Jeeps also have windshield catches bolted to the hood to secure the windshield when it is folded down.

Removing the hood is a simple process. Once you unhook the two front hood catches, simply lift the hood to the proper angle so that the pin can slip out of the hinge. It can be done by one person, but you are apt to dent or scratch the cowl, hood, or fender. With a helper, simply lift the hood and walk it forward.

This is a slip-style hood hinge for CJ-5s, CJ-6s, and M38-A1s popular from 1953 to mid-1970. Later Jeeps do not have this hinge setup and you must completely unbolt the hinge to remove the hood.

Hood catches come in many configurations. The traditional CJ-5 and many other models have one hood catch on either side of the hood attached to the fender. On the military MBs and GPWs, there is a hood catch and windshield catch.

Front Fenders

The 1946–1964 flat-fender civilian Jeeps have two bolts that fasten the front fender to the grille and two more that attach the little step behind the fender. There may be a brace from the fender to the frame. Finally, three bolts attach the fender to the main body tub. The World War II military versions are similar but have different brackets.

The front fenders on later Jeep CJ models are attached by three bolts from the fender to the grille and four more bolts from the fender to the main body tub. Look for a variety of braces, especially for the battery tray on the right side.

You'll also need to remove the hood catches from the top of the fenders. These Jeeps may also have a windshield washer fluid bottle, horn, and other items bolted to the inner fender. They need to be disconnected or removed too. The 1976–1986 CJ models have rubber fender flares that need to be removed before the fenders are painted. Replacement fenders and fender flares are readily available if needed.

The purpose of the catches was as follows: one set to hold the hood to the fender and the other to hold the windshield down so it does not bounce around while the Jeep is in motion.

Removing the front fender on a CJ is as simple as backing out three or four bolts from the firewall.

The grille attaches to the front fenders in a similar way. There are a few bolts on each side.

This bracket in the front fender well is not found on all models, but if your Jeep has them, they should be bolted to the frame. These bolts are usually very rusty due to their location, and they often break when you try to remove them. Do your best to remove any broken bolts and replace them.

There are two grille support rods from the top of the grille to the firewall on M38-A1s and CJ-5, CJ-6, CJ-7, and CJ-8 models. These bars prevent the grille and radiator from rattling, thereby preventing damage and adding strength to the front clip of the body.

Grille Support Rods

In early military and CJ flat-fender Jeeps, there is only one support rod running from the firewall to the top of the radiator. In the later CJ models, there are two grille support rods running from the firewall to the top of the grille shell. Remove these before removing the radiator or grille.

Grille

All civilian Jeep grilles are bolted to the chassis by a bolt in the center of it. The large bolt in the center of the grille holds the grille to the chassis. With the fenders removed and the center grille bolt out, unbolt the radiator from the grille. Then, disconnect the headlight and turn-signal light wires.

Once the grille is out, remove the headlights and turn-signal lights. Also, the welting along the top of the grille can be removed.

Willys MB, Ford GPW, CJ-2A, CJ-3A, and CJ-3B models use the same technique, except that they have a mounting pad and bolt to the frame at the outer edge of the grille in addition to the center bolt.

In the earlier Willys M38, CJ-2A, CJ-3A, and CJ-3B models from 1941 to 1964, there is only a single support rod that goes from the firewall to the top of the radiator.

There is a large bolt that holds the bottom of the grille to the chassis. Check to be sure that the rubber body mount and washer are still present under the chassis bracket. If not, the grille will be loose and squeak.

The World War II Jeeps and later CJ-2As, CJ-3As, and CJ-3Bs had two frame-mounted brackets that bolt to the grille bolts.

Floor Covers

In most of the flat-fender Jeeps, there is only a floor cover around the shifter. In the M38, M38-A1, and all the later CJ models, the floor cover in the center of the transmission tunnel comes out. This provides access to the transmission and all the bell-housing bolts.

Even if you are not removing the body, you have to remove the floor covers to get to the bellhousing bolts to remove the engine. Please note that the upper floorpan on the firewall may have a few bolts behind the heater ductwork. You may have to remove the heater to take out the upper floorpan.

Remove the floor covers from around the transmission and transfer

Alternatively, you can leave the radiator and fan shroud attached to the grille, but you need to drain and disconnect the upper and lower radiator hoses. Then, remove the grille as described above.

As previously mentioned, the M38 and M38-A1 models have hinged grilles that simply tilt forward for engine access. The hinges can be unbolted for grille removal. With the front clip removed, move on to the main body tub.

To save time and not damage the radiator and grille, remove them as one unit. Drain the radiator and then disconnect the upper and lower hoses, the grille support rod(s), and the three fender bolts on each side of the grille. Next, disconnect the wiring from the junction bar on the driver-side fender. Last, remove the big body-mount bolt on the bottom of the grille, and the radiator will lift out with the grille.

From 1941 to 1953, there was only a small removable floor cover around the shifter. By removing this, you only had access to the six bolts holding the shifter cover on top of the transmission. It wasn't much help in accessing the bellhousing bolts. Removing the shifter makes it easier to drop the transmission out the bottom.

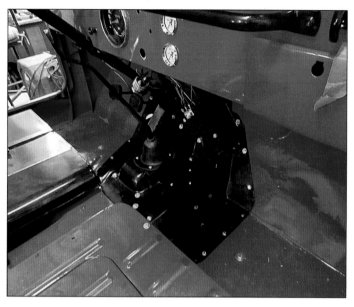

The manufacturer eventually provided full access to the back of the engine and to the transmission by installing large removable floor covers. This made it easier for both maintenance and engine removal.

fer case the same way and take it off with the body.

If your engine is still in place, unhook the choke and throttle cables from the carburetor. Replace them if you're doing a complete restoration. As they get older, they start to hang up.

I always cut the parking brake cable and install a new one once the body is back on. If you're not removing the body, replace the parking brake cable after the paint job. In both the flat-fender and round-fender Jeeps, they disconnect and attach in the same way. They didn't change much in 45 years.

case shifters and unbolt the steering column from the dash. Start undoing all the cables and wiring under the dash. If you are putting in a new wiring harness, just start cutting the old wiring anywhere that it is convenient. If you are saving the old harness to use again, label the wires and disconnect them carefully.

Cables

Removing the speedometer cable is easy. Just remove the large nut on the back of the speedometer. If it's on too tight, use a pair of tongue-and-groove pliers for leverage. You can disconnect the other end of the speedometer cable from the trans-

Steering Wheel

Removing the steering wheel is not as easy as it sounds. It's been on the vehicle for decades in good and bad weather. First, take the horn nut off. Be careful not to damage the threads on the end of the steering column after the nut is removed. I

Removing the steering wheel can be challenging. If you have tried pulling it, twisting it, and hitting it with a hammer and it still doesn't come off, the next step is to try a wheel puller. Hook the wheel puller to the underside of the steering wheel and put the center pin from the wheel puller where the horn button would go. Place an old nut or a flat piece of steel on the threads so that you do not damage the threads or bend the shaft. Use an air wrench (if possible) to crank down the wheel puller.

If you've tried everything else and you still can't remove the steering wheel, don't feel bad; this is very common. If you're doing a total restoration, you're probably committed to replacing the steering wheel with a new one. You can use a reciprocating saw to remove it. Make a cut on the outside edge toward the shaft. Slowly progress toward the steering column, staying away from the main steering shaft or threads where the horn nut goes. If done correctly, one or maybe two cuts will do it.

always replace the nut to cover the threads. This protects them while you try to remove the steering wheel. If you have to cut the old wheel off, be careful not to cut the threads.

Next, stand up in the Jeep where the gas tank was. Put your hands on the steering wheel at about 9 and 3 o'clock. Rock the steering wheel back and forth while pulling up toward you. Then, try the 12 and 6 o'clock positions and back again to 9 and 3 while pulling up. It may or may not come free at this point.

If it did not come loose, rent a steering-wheel puller. Be careful not to damage the threads next to the steering wheel where the horn button was. If the wheel puller did not work and tapping the back of the steering wheel with a big hammer on the back side of it didn't work either, you may have to accept the fact that you're going to have to buy a new steering wheel.

If you have tried all of the above a few times and it still did not come off, there is only one thing left to do:

use a reciprocating saw. Put a metal cutting blade in the saw and cut a notch in the same direction as the shaft of the steering column. Again, be careful not to hit the threads on the end of the column. As you're cutting, just as you get close to the steering column, you will hear a snap or pop and the wheel should be free and come off.

With the steering wheel removed, it's time to unbolt the steering column from the steering box and pull it up through the firewall so that the body can be unbolted and removed later. Check for any wires still attached to the steering column for the horn button and ignition or turn signal switch.

Body Mounts

There are not as many bolts holding the main body tub to the chassis as you may think. Depending on the model, two to four bolts along the rear crossmember bolt to

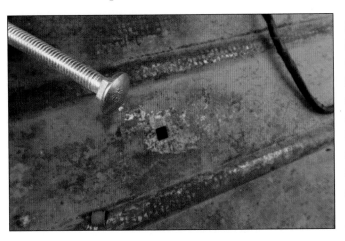

Situated on the floor in the cargo area, the round head of the carriage bolt allows cargo and boxes to slide freely without getting hung up on a hex-head bolt. This is the only place on the Jeep that carriage bolts are used.

Utilizing your helpers, walk the body over to your pre-setup sawhorses and set it on them. Give the body a little bit of a wiggle to ensure that it's safe and steady on the sawhorses.

the bottom of the body. On the rear floor between the rear wheel wells, there are a few carriage bolts (bolts set in a square hole with a round head).

On both the passenger- and driver-side floorboards, there is another bolt, and on both sides at the point where the front fender meets the firewall, there is a bracket that comes down from the bottom of the tub and bolts into the side of the chassis. On some older CJ models, there is a bolt inside the toolbox.

Don't be surprised if the old body mounting bolts are rusty and hard to remove. Cut them off if you like and replace them with new ones. On the lower front of the body, under the driver's and passenger's feet, there are bolts on each side that go from a body bracket into the side of the frame rail. There is a threaded hole

on the side of each frame rail. If you break one of those rusty bolts, it's not the end of the world. After the body is off, drill out the broken bolts and re-tap the holes. Some people prefer to drill all the way through and just put a new bolt and nut through the frame rail.

Main Body Tub

Once all the body-mount bolts are removed and you've checked for any wiring, cables, brake lines, and fuel lines, remove the main body tub. It takes four strong people to lift the body off the Jeep. If you have an overhead chain hoist, it makes life easier. Set up a few sawhorses and place the body on them. Place a few 2x6-inch wooden boards across the sawhorses. Turn the tub over to work on the bottom when you need to.

At this point, you have a rolling chassis with the drivetrain and axles intact. Access to the engine, transmission, transfer case, brake system, steering, and more is easier.

Final Components

Next, remove the engine, transmission, transfer case, and driveshafts one at a time. This is pretty straightforward with only a few bolts involved for each component. Obviously, these items are heavy, and having a strong friend is important. Set these aside and rebuild them as needed. Separate chapters in this book cover each component in detail.

Then, remove the wheels, tires, axles, and leaf springs. At this point, clean and repair the frame (chassis) and repaint it.

CHASSIS AND SUSPENSION

With regard to any restoration, it's paramount that all components of your Jeep are analyzed and examined for defects, swapped parts, and wear and tear. Being that these are Jeeps, most have been used for off-roading purposes and subjected to some abuse.

The good thing about a Jeep's chassis and suspension is that not much changed year to year, so these applications are all pretty standard.

Repairing and Refinishing

In Chapter 2, the chassis specifications (size and configuration) are listed for each model. It's always best to start with a straight chassis with no rust. But if all you have is one that needs some repair, it can be fixed.

Leaf Springs

If a leaf spring hanger on the frame is bent or broken, put the Jeep on a hydraulic car lift or on four jack stands. Make sure it's as level as you can get it. Unbolt the leaf spring that is affected and use an angle grinder with a circular cutting blade to cut the weld holding the spring hanger to the frame. You can use a cutting

torch also, but be careful not to cut into the chassis, just the weld.

If necessary, cut off the shackle and then use the grinder to remove the balance of the weld. If you can't find a new hanger, cut one off a junk Jeep. Clamp it onto the chassis in the same place as the one you took off and weld. Reinstall the leaf spring, and you're done.

Next, address the leaf springs. They may be in good condition, or they may need to be replaced. Replacements are readily available. Make sure to count the number of leaves before ordering.

You definitely want new bushings for the leaf eyes and main spring hangers at the frame. Having heavy-duty spring shackles with greaseable bolts lengthens the life of the suspension. Rubber factory-style bushings ride better. Polyurethane bushings are stiffer but more durable. It's your choice.

Buy new U-bolts to attach the axles to the leaf springs, and it may be a good idea to replace the centering pin for each spring. It's also a good time to replace the shim wedges used to rotate the axle and set the driveshaft angle. These tend to break over time.

Chassis

You can sandblast, scrape, or wire brush a rusty chassis, but the first thing to do is to pressure wash it. Once the dirt and grease that have built up over the years has been removed, you can see what you're working with. Remove any old muffler brackets, wiring, and fuel lines. Inspect the brake lines for rust. Once you have inspected the chassis and found no major rust, move on to cleaning.

If you found large amounts of rust and places that have rusted through, you may want to search for another chassis in better condition.

If you look closely, you might find places where the chassis was repaired. Look for steel plates welded on the top or side of the rails. Sometimes, previous owners have done a great job, but if you're not sure, show it to an experienced welder and get his or her opinion.

If the chassis is in good order and all repairs have been done, move on to painting.

It's best to use a heavy-duty chassis paint in a gloss or flat finish. The chassis can be spray painted or painted with a brush. It's easier to paint the chassis before installing the brake lines, fuel lines, and shocks.

This Jeep had a body-off restoration about 10 years ago. From the front, it looks level.

From the rear, you don't have to look hard to see that it is not level. In fact, it's off by more than an inch.

A torpedo level on the bottom of the bumper shows the problem.

The torpedo level shows that the body is leaning too.

Front or Rear Bumper

The front bumper is more prone to getting bent from hitting things than the rear bumper. The rear bumper often has rust that built up due to the rear tires kicking up mud and snow.

The front bumper is easy to change on any model. Remove a few bolts, and it comes off. On the early Jeep models, upper and lower plates on the end of the chassis are available, and they look good if you change them at the same time.

Many older Jeeps came with no rear bumper or with optional "bumperettes." These bumperettes are two oval metal hoops bolted to the rear crossmember.

I have seen rear bumpers bolted, riveted, and welded to the frame. On some of the models, you have to remove the draw bar (trailer hitch) to change the rear bumper. If your bumpers are in perfect condition, leave them alone. If not, it makes for a better finished product if you replace them.

Frame

After years of bouncing over bumps and holes, it's common to find even nice Jeeps with a few bends in the frame.

Various factors can cause the frame to be out of line besides abuse from traveling over rough terrain. You can also bend the chassis from abuse with snowplows and winches, etc.

Do not despair, it is not the end of the road if your Jeep has a damaged chassis. First, make any repair to the chassis that is necessary. Weld up the cracks and drilled holes. Replace the leaf springs, shocks, and bumpers if necessary.

When all repairs have been completed, take it to an auto body shop that has a frame straightening machine. The shop will make your Jeep chassis straight again.

Analyzing and Repairing Chassis Damage

1 *Use a tape measure on a level floor to check other dimensions.*

2 *Check the left and right sides and write down the measurements. Make sure the shocks are moving and not seized. The measurements on each side should be equal.*

There are many causes for chassis damage. This was a hole that was cut into the chassis to mount a snowplow frame. After years of abuse, it damaged both sides of the frame.

3 Use an angle grinder with a circular cutting blade to cut out all the damage. You can use a cutting torch, but the angle grinder makes a neater cut.

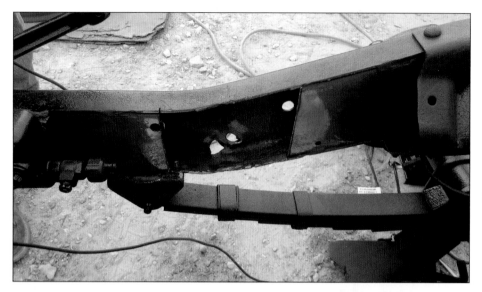

4 *After the damage has been removed and a new piece has been installed, repair the outside of the frame. Do not do both sides at the same time, otherwise it weakens the frame's structure.*

5 *Cut out a steel patch panel that is the same thickness as the frame and use a wire-fed MIG welder to replace the damaged area.*

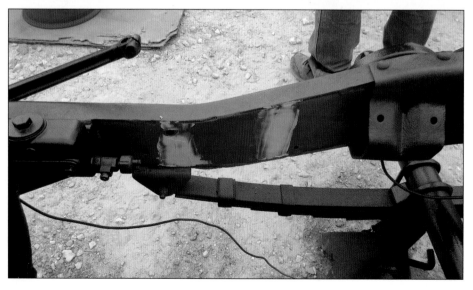

6 *Once the new plate is welded to the chassis, any extra weld needs to be ground off to make it look smooth. A little body filler is added for cosmetic purposes only. Once it is sanded down and painted black, you could never tell this chassis had damage.*

Ball Joints

It is pretty common to find worn-out ball joints on vintage Jeeps. This causes uneven tire wear. Ball joints are part of the suspension and allow the front tires to pivot and turn smoothly. It's best to replace them when rebuilding the front axle. There are a few specialized tools that make this process go easier. If a ball joint breaks, that wheel is no longer attached to the steering.

Shocks

There are only two types of shock mounts. One type has a threaded stud with a nut. The other has a smooth shaft with a hole for a cotter pin. Some restorers like to install the shocks after the chassis and body are painted so they don't get overspray on them.

Make sure the rubber bushings are in place in the shock eyes before installing the shocks. Some bushings are one piece and need to be pressed in while others have two halves. Putting some grease on the inside of the bushings will help them slide onto the mount and reduce any squeaking noises.

Shock absorbers are very important to the smooth ride of the Jeep and also for the comfort of the passengers. Matching the proper shock to your Jeep is very important. The wrong shocks can ruin the handling and the Jeep's ride.

High-performance gas shocks can have fancy names and purposes but may not do the job. Make sure that you stay close to the stock shock that was engineered for your Jeep. Also, with hydraulic shocks, you can see if fluid is leaking.

In some cases, people think they have a broken or weak leaf spring because the Jeep is leaning. Sometimes it turns out to be just a frozen shock stuck in the down position. Shocks are inexpensive and easy enough to change that if you have the least bit of doubt about their condition just replace all four and start fresh.

Removing the Shocks

1 *Before you attempt to remove the shocks, squirt some penetrating oil on all of the nuts. Between the rust and road grit, the threads are usually dirty.*

2 *After the nuts and cotter pins are removed, use a flat bar or large screwdriver to remove the shocks. Sometimes they come off easily. If they give you a hard time, use a larger pry bar.*

3 *Before installing the new shock, clean the shock mount. A wire brush, steel wool, or emery paper will do. After you clean it, put a dab of grease or oil on the mounting pin. This makes the new shock go on easier and prevent squeaking. Use a rubber mallet to tap the new shocks onto the mounts.*

4 *If you have a problem threading on the nuts, pull out your thread-chasing set. If you don't have enough room to use the handle, use a pair of adjustable pliers. Most of the time, you only have to refresh the first two or three threads.*

5 *It's good practice to use a new washer and cotter pin. You have gone to so much trouble to make this look good, just use new hardware whenever possible.*

6 *It is so much easier to replace some of the components under the body with it removed. Having access from both the top and bottom makes tasks like running brake lines so much easier.*

7 *Most shocks come in only a few basic colors. You can paint them any color you like depending on the color of the Jeep. They can contrast or match the Jeep's color. Just wipe them down with steel wool and spray them with a spray can.*

Grease

In the 1970s movie *Grease*, the script said, "Grease for Peace," and that is just what it'll give you: peace of mind knowing that your parts will not dry out, break, or fall off.

It is very important on vehicles such as Jeeps to keep them well-greased because of the nature of this vehicle being driven off-road through snow and mud. It's exposed to harsh elements. A film of grease separates parts and allows them to move. Ball joints, tie-rod ends, and other suspension and steering parts

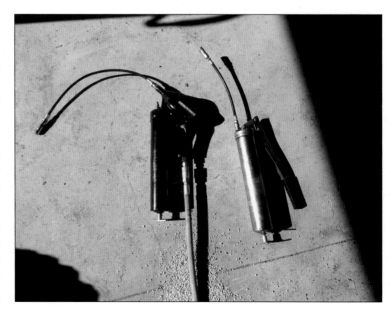

Grease guns are not very expensive. You can use a hand gun or an air-powered grease gun.

need grease to function. Without it, they seize or wear out quickly.

One of the most overlooked parts that must be greased is the U-joints on both driveshafts. It's very common for the grease fitting on the U-joint to be broken off. Check these often and replace them when needed.

This grease fitting is bad. It is filled with grit and has to be replaced. You can see the grease around the fitting, but it did not travel into the tie-rod end.

This is the way it should work. The grease should travel through the part and ooze out around the bushing.

Many of the older Jeeps have been running over rough ground for many years. The grease fittings take a beating. Get yourself a kit of replacement grease fittings. These are sometimes called zerk fittings.

Use an open-end wrench to remove the old fitting. Install the new one, and you're ready to go.

It's okay to use your hand to add grease to the steering ball on a closed-knuckle front axle.

This grease fitting is located on the end of the driveshaft near the yoke that holds the U-joint. In older Jeeps, this grease fitting is easily overlooked because it is commonly broken off.

This grease fitting is located in the U-joint itself. When you purchase new U-joints, sometimes the grease fitting does not come with it. If that's the case, buy the correct grease fitting and thread it into the hole.

VIN and Title

Some VIN tags are so deteriorated that they are held on by only one or two screws. The serial number tags may be missing after bodies have been changed or removed during restoration. The good news is that there are replacement plates that you can purchase for 1971-or-older models. Also, if you have a title to the Jeep, you have the VIN. Take a blank replacement dash tag to any trophy shop, and it can engrave or stamp the numbers onto the plate.

This is common when you're making one good Jeep out of two or three "parts" Jeeps. You'll have a choice as to which title to use on the finished Jeep. Leave the VIN plates on the bodies of the other parts Jeeps.

Law enforcement does not like to see a vehicle without a VIN. Remember that when you go to purchase a Jeep. Don't pay for it until you see the title, registration, and a bill of sale. Each state has its own laws about getting new titles when they are lost.

Some states won't issue a title for older vehicles. Military vehicles may never have had a title issued.

You always hear someone ask about a classic car, "Does it have matching numbers?"

In the case of the Jeep, there is no such thing.

All Jeep engines have a serial number. The engines were produced in higher quantities than the bodies and were used for agriculture and industrial equipment. The engine serial numbers do not

The metal serial number tag on Ford GPWs and Willys MBs built during World War II is riveted to the left frame rail on the inside just behind the front bumper.

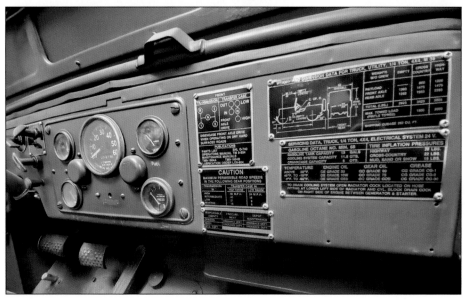

The brass serial number plate can be found on the right side of the dash in M38s and M38-A1s.

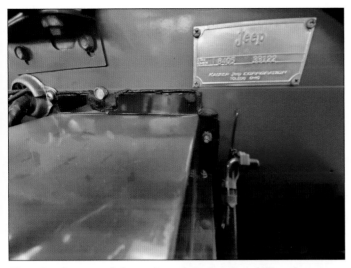

The aluminum serial number plate for the civilian Jeeps (CJs) can be found on the right side of the firewall.

The front crossmember on the Ford GPW frame is a flat channel. It is visible behind the grille.

correlate (match) to the chassis or body number.

Serial numbers can be found on the 1941–1945 Willys MB and Ford GPW chassis. There is a metal tag riveted onto the inside of the left frame rail behind the front bumper. It's on both the MB and the GPW.

On the military M38 and the M38-A1 models, the VIN is on the data plate screwed to the passenger's side of the dashboard.

On all the other CJs, it is under the hood screwed to the firewall. On most of the early CJs, it's on the passenger's side, and on the later CJs, it is on the driver's side. Over the years, some owners have moved the VIN plate.

It is important that the title and registration documents match the Jeep. Over the years, bodies, engines, and chassis have been removed and swapped to make one nice Jeep. The primary difference between a Willys MB and Ford GPW is that the front crossmember is different. Under the radiator (behind the grille) is a crossmember. In the Willys MB, it is an arched, round tubular front crossmember. In the Ford GPW, it is a flat U-channel front crossmember. When purchasing either one of the above, look at the papers and check the crossmember to see if it matches.

It will not make a difference in building your Jeep or in driving it. However, when the time comes to sell it and the buyer is looking at the crossmember, you could lose the sale if the chassis and documents don't match.

The front crossmember on the Willys MB frame is rounded and arched. It is visible behind the grille.

DIFFERENTIALS, BRAKES, WHEELS, AND TIRES

You can drive the most rusted-out Jeep with holes in the floors and fenders, but you're not going anywhere without adequate propulsion and stopping power.

Getting your Jeep from point A to point B requires a full overhaul of the driveline components. Books are available in the marketplace for rebuilding a Go Devil or Hurricane engine. So here, we'll be looking at rebuilding your differential and brakes.

Differentials

Some people refer to the differential as the "rear end" or "pumpkin." The ring, pinion, and spider gears are mounted inside the differential and are about the size of a small bowling ball.

The function of the differential is to transfer the power coming from the engine through the transmission and transfer case and down the driveshaft. The rotating driveshaft turns the pinion gear inside the differential, which turns the ring gear, which turns the spider gears (and side gears), which turns the axle

shafts, and finally turns the wheels.

Standard "open" differentials allow the Jeep's tires to turn at different speeds as it makes turns without binding. When the Jeep is going straight, all four tires turn at the same speed.

Not much goes wrong with the differentials, but when it does, it's usually catastrophic. The front axle's

differential repair is one of the more common repairs. In our example, three teeth have broken off the ring gear and are preventing the front differential from turning. The entire front axle needs to be disassembled. As with many labor-intensive repairs, once you have taken everything apart, you should replace the bearings, seals, fluids, etc.

Rebuilding the Front Differential

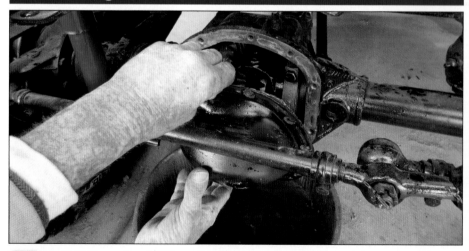

1 *Rebuilding a front differential may sound complicated, but once you learn the procedure it is rather simple. First, drain the gear oil from the differential by removing the lower drain plug. Then, remove the cover plate. This differential required repair because it had three broken ring-gear teeth, and the pieces were jamming the gears and not allowing them to turn.*

2 Once the differential cover is removed, you can get a good look at the ring and pinion gears. However, you will not be able to see the condition of the bearings. As with most labor-intensive repairs, once you've gone this far, it's always a good idea to replace as many wear items as you can, including the bearings. Once rebuilt, it should offer years of additional use.

3 To remove the ring and pinion gears, first remove the drive-shaft. This provides an opportunity to check the universal joint and replace it, if necessary. The driveshaft comes off by removing four small nuts from the back side of the pinion yoke and the two U-bolts. Keep them in a safe place and try not to lose them. A magnetic tray works well for this purpose. It's also a good idea to wrap the driveshaft U-joints with duct tape to keep the U-joint caps in place.

4 If you have manual locking hubs, remove them. Use a pair of snap-ring pliers, and a $2^{1}/_{16}$-inch socket (1/2-inch drive) for the spindle nut. Your service manual should cover the removal procedure and other tools. If you don't have manual locking hubs, remove the metal dust cap at the end of the axle. In some cases, you will find a snap ring. If so, remove the snap ring with a pair of snap-ring pliers. Front axles will vary through the years, and some may not have snap rings. Instead, they may have small nuts. If so, remove them.

5 *Remove the side bolts from the perimeter of the hub. It's okay to use air tools for this procedure.*

6 *Slide the hub off the axle shaft. This exposes a grease-filled cavity with a large spindle nut.*

7 *You will find two large nuts threaded onto a housing. Sandwiched between the nuts is a large washer. In many cases, the large washer has a tab or a lip that is bent over the outside nut to prevent it from turning. Simply bend that tab upward to keep the nut from turning.*

8 The spindle nut can be removed with a 2 $\frac{1}{16}$ -inch spindle socket. These are readily available online and at many auto parts stores.

9 After the spindle nut has been removed, remove the large washer. You will see a tab that fits in a slot, so it can only be reinstalled one way. The tab keeps the washer from turning so that when you bend the outer tab against the nut, everything is locked in place. After the large washer is removed, remove the inner nut.

10 The spindle nuts and washer may be damaged by people who didn't have the right tools. If that's the case, get new parts. Keep the nuts and washers in order and store them in a safe place.

11 *After the nuts have been removed, the entire hub should slide off. This will expose the axle shaft and the inner bearing. Inspect them for wear.*

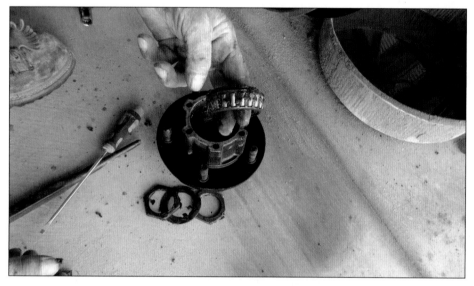

12 *When removing the hub and housing, inspect the bearings for damage. If you find damage, replace the bearings. Always repack the bearings with fresh grease.*

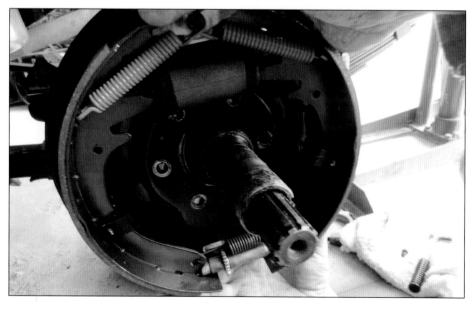

13 *By removing the six nuts from the backing plate, the entire brake assembly is easy to remove. You can now disconnect the steel brake line from the wheel cylinder.*

14 After the brake line is disconnected from the wheel cylinder, the entire brake drum assembly can be pulled off.

15 Then, the axle shaft should pull out. Simply wrap a rag around it for a better grip and pull straight.

16 The left- and right-side axle shafts are both assembled in the same way. One is longer than the other, which makes it easy to distinguish which side they go in.

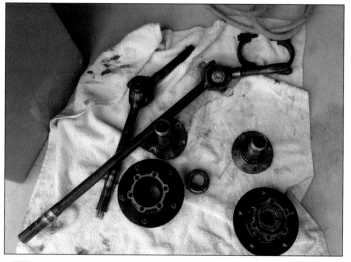

17 *Lay out your parts in a clean, dry environment so that they are ready to reassemble after you finish with the repairs.*

18 *This front axle was disassembled because of these three broken teeth from the ring gear. The broken teeth were causing the front axle to lock up and not turn. A great deal of labor goes into taking the front axle apart and putting it back together again.*

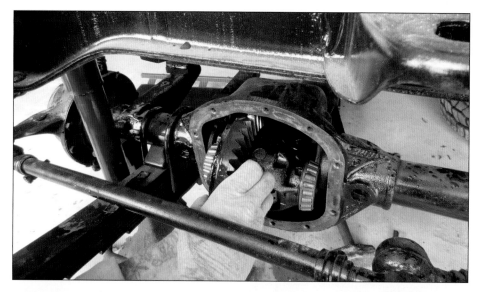

19 *After removing the axle shafts, remove and replace the ring and pinion gears. The ring gear is attached to a carrier that has the spider gears and side gears inside. Pull out the carrier to get access to the pinion gear. You may need a pry bar to pry the carrier out.*

20 *After the carrier is removed, reach inside and pull the pinion gear out. The ring and pinion gears should be replaced as a matched set. These gears mesh and need to be broken in gently for the first 500 miles. Please note that the front axle's ring and pinion gear ratio must match the rear ratio.*

Inspecting and replacing steering components is pretty straightforward. Wiggle them to check for excessive play and look for cracks or trail damage. Pump fresh grease into the zerk fittings to reduce future wear. Loose or worn steering parts can create unexpected tire shake and even death wobble, which is a real condition that is very dangerous.

gears with the original gear ratio so that the front and the rear differentials have the same gear ratio.

Your life and the lives of those around you depend on safe, reliable steering. When you rebuild a Jeep, it is a good time to inspect your steering components and decide what needs to be replaced. All of these parts are pretty affordable and cost less than your insurance deductible or a good lawyer.

Common wear items in the steering system are the steering box, bell crank, tie-rod ends, and drag links. It's also a good idea to inspect the pitman arm for cracks. The frame can also crack where the steering box mounts to it. There is a lot of leverage being applied to the frame in this area.

Brakes

The ratio of the ring-and-pinion gear is sometimes stamped on a tag that is bolted to the differential cover. If not, it is definitely stamped on the ring gear inside the differential. It is very important to order replacement

After the engine and the transmission, the most complex system in your Jeep is the brakes. Obviously, if

A cart on wheels serves as a workstation. All of the hardware is laid out in an organized fashion for quick access. Rebuilding your brakes is so much easier when you have a game plan. Take a few minutes before the project to organize everything.

These are four new wheel cylinders. If you get your new parts and they come with two large and two smaller wheel cylinders, don't be concerned because some Jeeps need them. Check your service manual and catalogs to see if you have the correct parts. It is not necessary to use new wheel cylinders. Many people like to rebuild the old ones using a rebuild kit. However, sometimes it is cheaper to buy new wheel cylinders. In the lower-right corner are new parking brake shoes.

New shoes and new brake drums are going into this old M38-A1. If you want to use the old brake drums, it's okay as long as they have enough metal in them to be turned down. Simply take them to a machine shop or brake shop, where they will put them on a lathe and grind down the inside of the drum to make it smooth. That way, there are new brake shoes matching up with a smooth surface of the brake drum. If you do not do this with the old drums, your brakes will make noise.

you put a lot of time and money into rebuilding the engine and transmission and have the vehicle running great, it would all be a waste of time if you couldn't stop the Jeep. From 1941–1975, not much changed in the drum-braking system used in these Jeeps. However, in 1976, front disc brakes became available to increase safety.

Ordering Brake Parts

By using the information from this book, identify the model of your Jeep so you can order the parts for your brakes. Keep in mind that in the 45-year span that I am covering here, the braking systems may have been changed. Although you have one model, you might have a different axle or braking system that was swapped in by a previous owner.

There are subtle differences in the drums of a CJ and military Jeeps (M38 and M38-A1), such as the three screws holding the brake drum to the axle. On some of the later Jeeps, the brake drum size has been increased, so before ordering parts, refer to your service manual for specific information regarding your model.

A popular brake upgrade swaps the original 9-inch brake drums for later 11-inch brake drums to improve the Jeep's stopping capabilities. Be sure to measure the size of the drum and shoes before ordering parts.

Rebuilding Drum Brakes

Here are a few of the tools for changing and adjusting the brakes. It's always better to have the correct tools. From left to right is the brake shoe adjuster, return spring spreader, hold-down spring remover/installer, and tool spring installer.

1 *First, get the Jeep off the ground. If you don't have a jack or jack stands, borrow or rent some. Use heavy-duty jacks stands for better safety.*

2 If possible, place the jack stand or blocks under the axle. If you do place them under the axle, make sure the stands do not touch the tires. If they do, you may have a problem getting the tire back on.

3 Another good place for the jack stands is under the axle in the center of the springs. Some people like to use two stands on each spring, placing them on either end of the spring. You can't be too safe.

4 Spray the lug nuts with penetrating oil and let them sit for a while. Remember that 1941–1971 military and civilian Jeeps have left-hand lug nuts on the vehicle's left side and the traditional right-hand lug nuts on the vehicle's right side. Try to break the lug nuts loose while the wheel is on the ground. That way, the wheel won't rotate when you apply torque with the lug wrench.

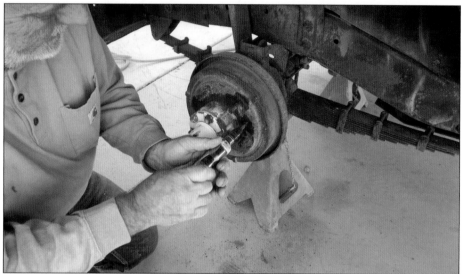

5 On military Jeeps, the drums are also attached by three screws. A large screwdriver should remove them. If you're unable to manually remove them, use an impact screwdriver (a heavy-duty screwdriver that you hit with a hammer while twisting the handle).

6 CJs and military drums are the same size. They are interchangeable. The only difference is that military Jeeps have three screws holding the drum to the hub. Some people have lost the screws and run the Jeep without them. If the screws do not come out or they break off, drill them out and re-tap the hole.

7 After years of use, the inside of the brakes gets dirty. Many have leaky wheel cylinders, brake shoe dust, and rust.

8 Remove the hold-down springs first and remove the upper horizontal spring. Then, the shoes should come off. Sometimes the shoes are stuck in the wheel cylinder, and you may have to pry them out. Don't worry about damaging the wheel cylinders if you are going to replace them.

9 *If you do not have a pressure washer, use a cleaning degreaser and a brush.*

10 *Thoroughly brush the backing plate. It's alright to reuse the small parts again (springs, hold-down pins, etc.) if they are in good condition. Clean and inspect them.*

11 *Get a can of black heat-resistance paint or black engine paint. This is done not only to make it look good but it also helps to keep it clean. Most people use black, but you can use any color.*

12 A small parts kit can be purchased for the brakes. If you do not have that, you can use the old parts if you clean and inspect them. You can use a tumbler with plastic media if you have one. If you purchase the tumbler from an auto parts store, it will usually cost more than if you buy one from a sporting goods store that sells ammunition. They use the same tumbler to clean brass for reloading, and you will save money.

13 Make sure to loosen the brake adjuster before you install the brake. If they do not work, you may have to take all the brakes off and replace the adjusters. You can get to the adjusters on the back of the brake backing plate. Loosen the nut and turn the adjuster with pliers.

14 After the backing plate is clean and all of the brake parts have been removed, replace the brake shoes. Install the lower horizontal spring and the left side of the upper spring. This makes it a lot easier to install.

Disc Brakes

Although you can have a machine shop turn the rotors on a lathe to make them smooth again, it is usually cheaper to replace them with new rotors. The same is true with calipers and pads. You can buy "loaded" calipers and simply bolt them on for less cost than trying to rebuild them.

Brake Lines

The brake lines are like the arteries in your body delivering blood to all of your extremities. If one of your brake lines is broken or leaking, you lose the fluid in the master cylinder and the Jeep will not stop. If one of the brake lines is crimped or pinched and no fluid is getting to a wheel cylinder, that wheel will not have working brakes, creating a tendency for the Jeep to pull to one side.

It is very common that after decades of road use that brake lines are rusty and porous and start to leak. They are not expensive to buy or make. You can purchase a premade set that you simply install. It is worth the extra cost to upgrade to stainless-steel tubing. Otherwise, you can make your own brake lines. All you need are a brake-line tube bender and a double-flaring tool with the proper end fittings.

A variety of clamps hold the brake lines to the rear differential and the chassis. You can save the ones from your Jeep if they're reusable or buy a new replacement set.

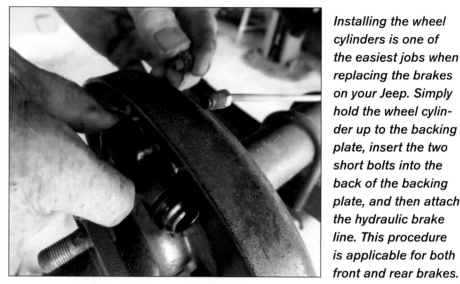

Installing the wheel cylinders is one of the easiest jobs when replacing the brakes on your Jeep. Simply hold the wheel cylinder up to the backing plate, insert the two short bolts into the back of the backing plate, and then attach the hydraulic brake line. This procedure is applicable for both front and rear brakes.

By this time, the wheel cylinder should be installed and fastened with two bolts from the back side of the plate.

Replace the round hold-down spring. Insert the pin from the back of the spring and the round washer from the front. You can use a tool or do this with your hands. Just twist the pin through the slotted hole. Notice how clean the spring and washer are? They were put through the tumbler and all the rust and dirt came off.

Replacing Brake Lines

1 If you made the decision to replace the brake lines, do not throw anything away until you're done. Remove all of the clamps and tie-downs that hold the lines to the Jeep. Remove all of the C-clips or horseshoe-looking clips with tongue-and-groove pliers, a screwdriver, and a hammer.

2 Some of the brake-line clamps bolt to the differentials. If the bolt breaks off, drill it out and re-tap the hole. Some people are afraid to drill this out with the fear of drilling into the differential housing. If this is the case for you, then don't drill it. Use a large clamp to hold the brass block and lines.

3 Some clamps strap to the differentials. Unless you are going for a purely stock restoration and want everything as original as possible, feel free to use custom-fit clamps whenever possible.

4 Remove the brass distribution blocks or fittings. I like to put them in a tumbler and shine them up, or you can purchase new ones. If you cannot get the old brake lines out of the brass blocks, cut the brake line, put the block in a vise, and use clamping pliers to remove them.

5 This connection is under the floor and connects the steel line to a rubber flex line. Four rubber flex lines connect to the steel lines, and most use a C-clamp connection.

5 Remove all items attached to this brass junction block atop the front axle. Note the orientation of each line by photographing everything before disassembling it.

6 Unthread or cut off the line on the rear backing plates above the rear axle. If they're not too rusty they will come off. If you can't get them loose, cut them off and replace everything.

7 Start by installing all of the rubber flex lines. Most of them attach with the use of horseshoe clips. Some ends of the rubber brake lines have a fitting that swivels. Others are fixed in place.

8 Next, install the brake-line clamps and sleeves on the differentials. Leave them loose until you install the new brake lines and then tighten them. This provides some wiggle room when making the lines fit.

9 Feed the pre-bent or custom-bent brake lines down the long runs, down the chassis, and along the differentials. Make sure the clamps are placed in the same position as they were prior to removal. If you want to add more hold downs or clamps, do so at this time. There is less rattling and and a lower chance of road damage if you do.

10 Unfasten the "S" brake-line fittings in preparation to replace it.

11 *The S-line installation is next. Remove the old unit. It's easier to do this before you install the front brakes, but you can still do it now. Loosen the bolts that hold the front wheel cylinder to the backing plate. Once they are loose, hand thread one end of the S-line into the back of the wheel cylinder.*

12 *Next, hand thread the other end of the S-line into the end of the rubber line on the chassis (don't put the C-clamp on yet).*

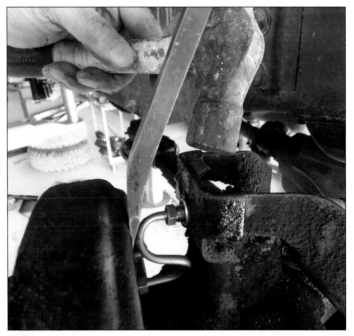

14 *Insert the C-clamp on the chassis side and slowly tighten the two bolts in the back of the wheel cylinder. You can use a screwdriver or another tool to keep the S-line in place while you gently tap on the clip with a hammer.*

13 *Now you have the loose S-line attached to both ends.*

Depending on the model, you may have three or four flexible brake hoses. It's always a good idea to replace the rubber brake hoses for each Jeep restoration. They tend to split and leak over time.

Brass fittings on the end of the brake lines usually do not rust. However, the steel brake lines inside the fitting can rust. If a leak occurs in this area, you might have to replace the entire brake line. It is not practical to cut off the fitting, so replace it and re-flare the end of the line because the line will be too short to reach its destination.

Master Cylinder

The master cylinder is the heart of the brake system, pumping fluid down the arteries to the wheel cylinders. The 1941–1965 Jeeps have the master cylinder mounted on the frame rail where it is exposed to the elements and collects lots of salt and mud. When removing and replacing the master cylinder, you might find it difficult to detach the brake lines due to rust. Notice the metal shield under this master cylinder. This is to protect it from the environment but mostly is a heat shield from the exhaust system. These shields are usually found on the M38 and M38-A1 military versions of the Jeep.

Access to the single-circuit master cylinder mounted beneath the floor is achieved through a round inspection plate on the driver-side floorpan. Once the floor plate is removed, you will see a large hex head on the filler plug. Once the filler plug is removed, use a funnel to refill the master cylinder. The whole process is simple but awkward. It is very easy to overfill the reservoir.

For almost 30 years, the same

Now that you have the Jeep stripped of all the old brake lines, it's time to replace them. The easy way is to purchase a new set. They come in a kit with all the ends made and with fittings. They work well and save time.

If you can't get the steel lines out of the brass fittings, use locking pliers. Throw the brass in a tumbler or use a wire brush and cleaner. If they are corroded or bent, buy new ones.

To fill the master cylinder with brake fluid, there is an approximate 5-inch hole in the floorboard with a corresponding cover plate. Depending on the year and model, this cover plate has four or five bolts. Beneath this floorboard, you will see the top of the master cylinder with a large hex head attached to a plug. Remove the plug and use a funnel to top off the fluid. Then, bleed the brakes at each wheel cylinder starting with the one farthest away.

You can see that the master cylinder is attached to the Jeep only by two bolts that go through the steel bracket of the master cylinder and into the chassis.

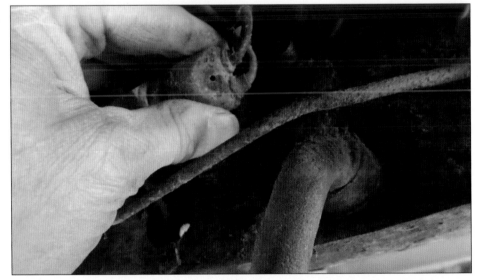

The first thing to remove is the wiring from the brake light switch. In this photo, it is a waterproof plug on an M38-A1. On a civilian CJ, it has two individual wires plugged into the switch.

In this brake job, we are not only replacing the master cylinder but also all the brake lines and brakes. It was easier to cut the brake lines at the master cylinder. If your brake lines are in good condition and are to be reused, you will have to unbolt them using the proper wrench.

We decided to reuse the brass distribution block for the two brake lines, so we put it in a vise to obtain a better grip when removing the parts. Do not overtighten the vise, as you can crush the brass and make the holes out-of-round.

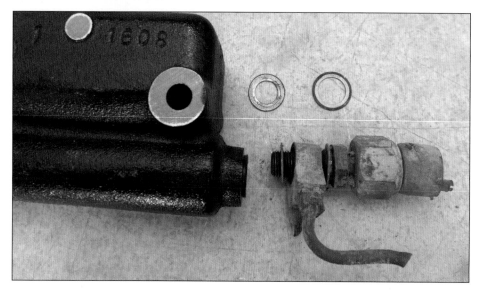

Next to the master cylinder are two copper washers. They have different-size holes and can only go on one way. One goes between the brass distribution block and master cylinder, and the other one goes between the distribution block and brake light switch.

This is a good example of all the parts assembled to the master cylinder. The main bolt holding this assembly to this cylinder hasn't been tightened yet. Wait until you have threaded in the brake lines so that you have flexibility to align everything.

At this time, the master cylinder is remounted to the chassis. Brake lines are also attached to the distribution block. Now, tighten the main bolt holding this unit together.

Connect the plunger into the brake pedal and plug in the brake light switch wires. If you have any leaks, it can only be from two things: the brake lines are not tight enough, or it is leaking around the copper washers if the main bolt is not tight enough. You may need to take it apart and add an extra copper washer to stop the leak.

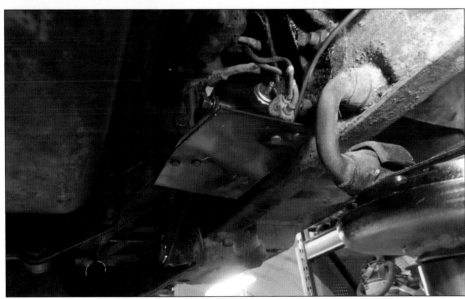

Last, the new heat shield is bolted back onto the frame rail.

The single master cylinder brake system was replaced with a safer dual master cylinder beginning in 1966. To fill the master cylinder with brake fluid, there is an approximate 5-inch hole in the floorboard with a corresponding cover plate. Depending on the year and model, this cover plate will have four or five bolts. Beneath this floorboard is the top of the master cylinder with a large hex head attached to a plug. Remove the plug and use a funnel to top off the fluid. Then, bleed the brakes at each wheel cylinder starting with the one farthest away.

type of master cylinder was used. It had a single circuit for all four brakes with only one reservoir. Any failure resulted in the total loss of brakes.

In 1967, the CJ-5 switched to a dual-circuit brake system with two reservoirs. This meant that if you had a brake line or hose fail on one axle, you could still stop by using the brakes on the other axle. In most cases, these master cylinders are interchangeable, so don't rely solely on the year and model of the Jeep. You need to physically match the new master cylinder to the current one before removing it from your Jeep.

Parking Brake

The old drum-style parking brake on the back of the transfer case found on 1943–1971 Jeeps is probably the most common item that does not work. If you have this style of parking brake taken off the Jeep for repair, this is the ideal time to change the rear seal on the transfer case.

To do so, take out the old seal with a seal puller, hammer and chisel, or pry bar. Make sure the seat is clean and wipe it with some grease or oil. (Chapter 9 covers this procedure in detail.)

The parking brake for the M38 and M38-A1 military model is different than the CJs. On the military Jeep, there are internal and external shoes that grab the drum to hold the Jeep. There is no cable to activate the military brake (just a ratchet lever between the seat with a rod). On CJs from the mid-1940s, a cable activates two internal shoes, much like the brakes on the wheels. The parking brake drum is vented to keep it cool.

The M38 and M38-A1 both have the parking brake handle mounted between the seats. They have a ratchet handle that doesn't wear out too often, but when it does, you can't fix it. It will need to be replaced.

Setting the Parking Brake

To set the parking brake, all 1941–1953 models have a mid-dashboard handle that is pulled toward the driver. Turning the handle releases the cable.

On the 1955–1971 CJ-5, the handle is under the dashboard to the left of the steering column. Pulling the handle straight out activates the parking brake; turning it releases it.

The 1972-and-later CJ models have a ratcheted foot pedal to activate the rear drum brakes and a release handle under the dash to release the cable. Parking brake cables are adjustable and can wear out or stretch over the years and need to be replaced, especially if the brake is not working. ■

The 1972-and-later CJs have a ratcheting foot pedal on the left side of the body tub. A black plastic handle above the pedal releases the parking brake.

The parking brake handle on 1941–1953 Jeeps is in the middle of the dash. On the 1955–1971 CJ-5s, it was moved left of the steering column and mounted under the dash. The 1972-and-later CJs have a ratcheting foot pedal.

The parking brake handle and cable (missing) on early MBs, GPWs, and CJs are found in the center of the dash. Pulling the lever mechanically pulls the parking brake shoes against the drum mounted to the back of the transfer case.

In the M38-A1, when you lift the handle lever between the seats, the two shoes come together and clamp down on the drum, holding it tight. There is no cable, just a handle with a rod that connects directly to the shoes.

The 1941–1943 military MBs and GPWs had an early parking brake with upper and lower shoes outside of the drum activated by a single cable.

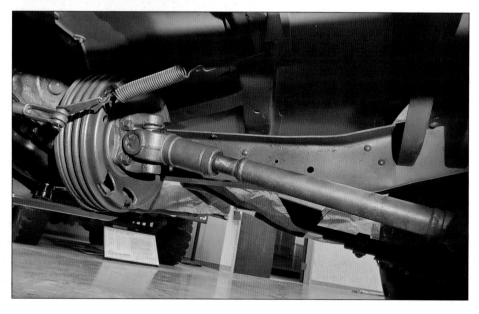

This is a typical CJ drum parking brake. It has a drum with internal shoes just like the brakes on the wheels with two shoes and an adjuster. This is an early mechanical brake. When you pull the dashboard handle, you pull a cable that pulls a lever, pulling the brakes outward, rubbing against the drum, and holding tight. When you turn the handle on the dash, a return spring under the Jeep releases the parking brake shoes.

The inside and outside shoes are shown with the lever.

Assemble as much as you can outside of the Jeep before installation. There is not much room to work under the Jeep when installing the parking brake.

Wheels

Many Jeeps have been customized by removing the original wheels and replacing them with wider aftermarket wheels. Unfortunately, most people did not keep their original wheels, and at this time, there is a shortage of them. Finding original wheels that are still straight can be a challenge. Sometimes, the holes for the wheel studs are wallowed out. If the wheel is still good, the cost to sandblast any rust or layers of old paint needs to be considered. New reproduction wheels are available and a lot easier to find and paint.

This is a typical Jeep wheel that has been on the road for 30 to 40 years in all kinds of weather. Of course, you can wire brush it or sand it by hand. The best procedure is to have it sandblasted. First, dismount the tire. Then, bring the wheel to a sand or media blaster. Sandblasting removes not only the surface rust but also allows you to see if there is any major metal deterioration on the wheel.

When you get the wheel back from the sandblaster, put a good coat of primer on it immediately. It's a good idea to prime both sides of the wheel. This will give you time to decide on what your finished color will be. Spray cans are more than acceptable for both primer and finished coats on your original-style steel wheels.

Non-directional tires are still available from several retailers. These tires are commonly associated with military Jeeps. They are not as good as a good pair of modern snow tires, but they will surely dress up any military Jeep.

These tires still had the labels on them. Although they looked new, they were 20 years old but never used. The tire company refused to mount them due to liability concerns. Tire manufacturers have been stamping date codes into the sidewall of each tire since 2000, so check the codes before wasting your time trying to mount any tire more than 10 years old.

Replacing Wheels

The 4½-inch center-hole bore has not changed in 45 years. This size allows manual locking hubs to pass through. The 5x5½–inch bolt pattern did not change either. However, there are a few variations in the wheel width. Vintage Jeep wheels vary from 4½- to 6½-inch width. The most popular is the 5¾-inch width. The width and diameter is stamped on the inside of the rim. Jeeps from 1941 to 1971 can use 600-16 or the 700-16 tires. The 15-inch wheels were standard for later CJs in 6-, 7-, and 8-inch widths.

Combat Rims

Combat rims were used on the 1941–1945 military Jeeps. They made it easy for the GIs to change and fix a flat in the field. After the flat tire and rim are removed from the Jeep, the same lug wrench is used to separate the front and back of the rim. There is a row of eight bolts, and when you remove the nuts on the back side, the rim splits in two and allows the tube to be removed and fixed. Then, the process is reversed. Tubes are required with combat rims. New combat rims are hard to find but are available. Combat rims provide great authenticity for a military restoration.

Lug Nuts

Lug nuts have not changed much over the years. Remember that 1941–1971 military and civilian Jeeps have left-hand lug nuts on the vehicle's left side and the traditional right-hand lug nuts on the vehicle's right side. You'll encounter the difference when you try to remove or install them.

Tires

If we start at the beginning with the World War II Army Jeep, most people think of the "Army tires." That is the Hollywood image of the Jeep that has been seen for 60 years in the movies and in photos. Many people call them Army tires, but today they are commonly known as non-directional tires. They used to be all over the place, and we had no problem getting them. Today, it's a different story.

If you go to most national tire franchises, they won't have them in stock, but they can order them for you. Non-directional tires are being reproduced and there are a few tread patterns.

If you are not restoring a military Jeep, you most likely need a more conventional tire. There are many choices. Many people want their Jeep to look as it did when it rolled off the assembly line. Once you have the correct wheels and know the tire size, you have many choices when it comes to tire tread.

There are many styles of street tires. They tend to be smooth riding, easy steering, and not very expensive.

At the other extreme, there is the knobby-tread off-road tire for gripping dirt and snow. They are great for traction but not for such a smooth ride on the street.

Somewhere between the street tire and the off-road tire is the all-terrain tire. It is called by many names and has many combinations of tread to give the driver what he or she needs. These are a great compromise for most Jeep owners.

You need to have a direction or goal in mind when starting this project. If you build your Jeep for parades or car shows, you want a more original tire. Some people have one set of wheels and tires to meet a particular goal, and then switch to another set when they change their mind.

Spare Tire

Most people consider the spare tire as part of a set. They always include the spare into the budget, and it looks nice if it matches the set on the Jeep. Spare tire carriers and other accessories are available on both the internet and in catalogs.

Make sure if the spare is mounted on the side that the bracket is secure and does not wobble. On 1949–1971 Jeeps, a small block of wood on the side of the Jeep takes up the space between the spare tire and the rear quarter panel to prevent this wobble. It is called a spare tire dampener.

Always check the lugs on the spare tire to make sure they are tight. Lug nut locks are also available if you need to put one on your spare.

Always be aware that the spare is hanging off the side of the Jeep. It's easy to forget and hit something with it.

The World War II military Jeeps didn't have a tailgate, so the spare tire was mounted on the back. For civilian use, most people find the tailgate more versatile, so the spare tire was moved to the side.

Many aftermarket wheels with various finishes can be installed on your Jeep. Many owners like to use stock rims on vintage Jeeps, but you have many choices and sizes from which to choose.

It is common to find old Jeeps with big, oversized tires on them. They are readily available, and some Jeep owners try to turn their Jeep into an off-road dune buggy. You'll find that the steering is greatly affected by using these wide tires. Also, wider tires may rub on the leaf springs or frame.

Depending on the climate where you live or the usage for your Jeep, there are a wide variety of tire tread patterns. The owner of this Jeep in South Carolina had the wheels sandblasted and powder coated. Then, off-road tires were mounted for both snow and off-road use.

ENGINE

For 30 years, the Jeep 4-cylinder engine remained relatively the same. From 1941 to 1953, there was the flat-head engine (also known as the L-head or Go Devil engine). This was basically 1930s technology. In the rush to deliver the Jeep for military use, this engine was adequate and delivered for the next 10 years.

Realizing that the Jeep needed more power and efficiency, engineers came up with the Hurricane engine, which was introduced in 1953. The F-head engine had four intake valves in the head and four exhaust valves in the block. This is called an "intake over exhaust" design.

AMC bought the Jeep brand in 1970 and began offering its own OHV engine in 1971. The lower part of the engine is very much the same, but the heads and the valves are different. The AMC engines offered more horsepower and efficiency.

Removing the Engine

Safety is the first and foremost factor when taking any vehicle apart. When removing the engine, you are going to be lifting approximately 400 pounds up and over the grille and placing it in a safe position to

work on. Keeping a neat workplace is very important for safety reasons. You don't want to drip antifreeze, water, oil, or transmission grease on the floor if it can be avoided. That's why planning the removal and disassembly of your engine is important.

Don't attempt to do this on your own. Even if you are fully capable of disassembling the vehicle, it's always a good idea to have a backup person just in case. Some people think that neatness doesn't count because they are doing a complete restoration and that they can cut all the lines and wiring. However, it's a good idea to take everything apart as if you are going to use it again because you may have to.

If you are doing a complete restoration, you may find it easier to remove the hood, fenders, and grille first. This provides more room to work, and you don't have to lift the engine as high to remove it. If you are doing a complete restoration, many people remove the entire body before approaching the engine and transmission.

When you get the rebuilt engine back from the machine shop, get set up to reassemble the accessories. Try not to do it on the floor. I like to use a

cart or table that is strong enough to hold the weight. I use my chain hoist to lift it onto the table and keep the chain hoist attached to the engine during the dressing process. Even though I block up the engine with wooden blocks, it's safer if it is still attached to the hoist. This puts the engine at eye level and makes things easy to see and work on.

An engine stand will keep the engine at about table height. You can rotate the engine from side to side for easy access. The only thing that does not work is that you cannot attach the flywheel and clutch while the engine is on the engine stand. Some engine stands are cheap and top-heavy. Be careful when rotating the engine. Be sure the base is wide enough so that the stand does not tip over.

Restoration Strategies

There are three ways to restore an engine: do a backyard overhaul, go with a long-block, or do a complete rebuild (you alone or along with a machine shop).

Backyard Overhaul

When we were kids and didn't have much money, we did a

backyard overhaul. You don't need many tools to do this, but you do need a honing tool, an electric drill, some valve-grinding compound, and a hand tool (a wooden handle with a suction cup on the end) to lap the valves.

Using an old valve-spring compressor and a ring compressor, we disconnected the rods from the crankshaft and removed the pistons. Then, we removed the old rings with a screwdriver; most of the time, they broke off in pieces. That did not matter because we were replacing them with new ones. We then cleaned the pistons with a wire brush and a scraper. It was very difficult to clean out the oil-ring grooves in the piston. Once we cleaned it, we had to muscle the new rings onto the old pistons. Once in a while, we would break one and have to buy another set.

Today, there are large ring pliers for removing and installing the rings. With the right tool, there is a lot less chance of breaking them. If we had a Jeep that smoked a little and had low oil pressure but it wasn't knocking, we did a backyard overhaul. Add new piston rings, crankshaft bearings, valve lapping, a new rear main seal and gaskets, and we were back in business. This was a patch-up job. It lasted for a while, but it was not a rebuilt engine.

Be careful when you purchase a Jeep and the seller tells you it has a rebuilt engine. Ask to see the invoice and worksheet from the machine shop. That will tell you what was done. If there is no supporting documentation, it's just a used engine. A lot more goes into rebuilding an engine properly so that it will give you years and miles of reliability.

When delivering an engine to a machine shop, remove all of the accessories beforehand. Leave all the sheet metal on (valve covers, oil pan, etc.). Tell them what you know about the engine. Point out any broken studs, such as manifold bolts, so they can be fixed.

The first thing the machine shop does is remove the head to see the condition of the pistons and valves and locate any cracks in the head or block. The shop can see if your engine had a blown head gasket or burnt valves. It might give you some answers as to why your engine doesn't run well. But most of this does not matter because this engine will now go through a complete rebuild.

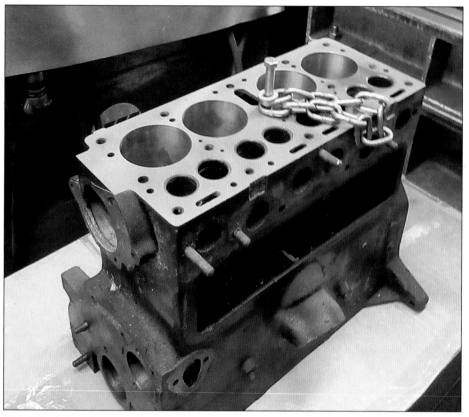

This block has been boiled out and cleaned. The head bolts have been removed, and the top of the block has been milled. Milling provides a flat surface. When the milled head is installed, you will have a perfect match, and the head gasket will seal well.

You can see the milled head that goes on this engine is clean and flat. All the water jackets are clean, and the flat surface will match up with the block.

Go with a Long-Block

With a short-block, you send your engine to a machine shop and it machines the lower block: boring the cylinders, installing new pistons, turning down the crankshaft journals, installing camshaft bearings, etc. You will get it back without the head installed or the sheet metal attached. You save money and you finish assembling it.

This is not very popular anymore. I learned a long time ago, "For 5 cents more you can go first class." You can't beat a professional at his or her own trade. Let them finish it.

Pay for a long-block. Then, when you get it back, it's under warranty and all you have to do is attach the accessories.

Complete Engine Rebuild

There are two ways to do a complete rebuild. The first way is to have the machine shop machine all of the parts that need machining and you assemble the engine. The second way is to have the machine shop machine everything and completely assemble it for you.

Either way, find a reliable shop. If there is more than one engine rebuild shop in your town or city, visit them. Talk to the owner and ask around to check the shop's reputation. Go to local shows and mention the builder by name and hear firsthand from customers. Find out what's included in the rebuild and if there's a warranty. Most shops offer 12 months or 12,000 miles. Compare prices and ask how long it will take.

Some local auto shops offer rebuilt or remanufactured engines from national companies that arrive in a crate and ready to go. These engines are built in massive facilities that specialize in this process.

They are unlikely to offer the older 4-cylinder engines but may have the later 1971-and-up AMC engines.

Working with a Machine Shop

Many people would like the experience of putting the engine together themselves. The challenge and the creative process can be very rewarding. When it's complete and running, you can step back with a great feeling of satisfaction knowing that you built it. That sounds good, but you can't do it alone. Unless you have $100,000 worth of machines, you can't turn the crankshaft, bore out the cylinders, replace the camshaft bearings, mill the block, head, and flywheel, etc. What you *can* do is work with a professional machine shop.

Rebuilding an Engine

1 Your engine may have been rebuilt once or twice during its lifetime. When you or the machine shop look at the top of the piston, you will see a number stamped on it. For example, "20" means that the piston is 0.20 inch over the standard-size piston that came from the factory. If the shop believes that the engine needs to be rebuilt again, it will replace a 0.20-over piston with a 0.30- or 0.40-over piston. The cylinder bore then has to be made larger to match. This boring machine can complete that task.

2 After the boring is complete, the next step is to use a honing machine. Three or four mounted stones on the tool rotate fast, removing any drilling marks or grooves. This leaves the inside of the cylinder with a mirrorlike finish. It puts a fine crosshatch pattern on the cylinder to retain a film of oil for the piston rings to slide on.

3 *This is the valve-spring tool. It is used to compress the valve spring so you can remove the two half-moon keepers that hold the spring to the valve stem. Do not lose the keepers. When putting them back on, use a little grease to hold them in place.*

4 *Once the valves are out, the machine shop inspects them to determine if they should be replaced or if they are good enough to be reground.*

5 *These are the valve lifters. The flat circular end rides on the cam lobes. The other end has the adjusters that push the valve stem up and down. The shop will advise you as to their condition.*

6 Consult the service manual for the proper gap when adjusting valves. Find the feeler gauge that matches that gap. Use two open-end wrenches to open or close the gap. Keep using the feeler gauge to slide through the gap. When you reach the correct gap, lock the opposing nuts.

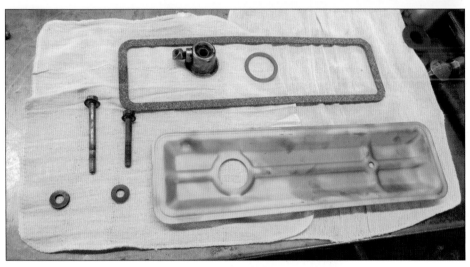

7 Make sure the side cover has the large hole facing forward. There are a few types of breather attachments on the side cover. Glue on the gasket and the side cover. You will need the longer bolt on the front side and the shorter one in the rear. On the F-head engine, the throttle linkage will be attached to this bolt.

8 These dirty, burnt pistons have some of the rings rusted in place and will be tossed. The machine shop will remove the pistons and check the pushrods.

9 Here are the connecting rods after being cleaned up and inspected for cracks, bends, or twists.

10 Once the connecting rods pass inspection, new wrist pins, pistons, and rings are assembled. They are ready to install.

11 The cylinders are bored to size with new pistons to match, but the installation is not complete until the clearance is measured with a feeler gauge. Once the machinist is satisfied that the clearance is acceptable and that the oil will flow properly, it goes to the next phase.

12 *Just as with the cylinders, the crankshaft must be ground to the next size smaller to remove any uneven wear. The machine shop puts it on a large metal lathe and uses a grinding wheel. It grinds each journal and makes sure they are the same size. Then, the shop matches the bearings to the new crankshaft size.*

13 *The crankshaft is in the engine and ready for the pistons. It's time to bolt the connecting rods to the crankshaft.*

14 *The crankshaft is held in with three main bearing caps. Each one has a new bearing in it. The rear crankshaft main bearing has a seal also. The oil pickup has been installed.*

15 *A new gasket goes on the block before bolting on the oil pump.*

16 *Make sure the oil pan is clean and that there are no holes or creases that may leak. You can clean it and knock out any dents. Some have skid plates on the bottom, many are welded on, and some are attached by pan bolts.*

17 *Make sure the machine shop has added the six flywheel bolts: four conventional and two tapered. If any are missing, the machine shop will have to remove the crankshaft to install the bolts. The flywheel can only be bolted on one way. The holes in the flywheel must match the six-bolt pattern.*

18 Connecting rod bearings come in a matched set. Half goes on the connecting rod and the other half goes on the bearing cap. Put a small amount of assembly lube on the bearings so that they're not dry when you start the engine.

19 Main bearings can only go in one way. This is also true for the rear main seal. Put some assembly lube or oil on the bearings and place them in their slots. Research the torque specs in advance and torque them down using a torque wrench.

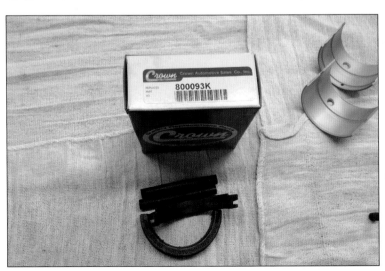

20 The rear main seal is very important in your engine to keep oil from dripping out. These seals may have two curved halves, or they may be two round pieces. Pay attention when installing them to ensure that they seat properly in the grooves. They are not expensive, and they often leak if not seated properly.

21 Sometimes, when tapping the piston down through the cylinder, the connecting rods have a tendency to hit the crankshaft. Be careful not to damage the new connecting rod journals on the crankshaft. After you have torqued down the connecting rod caps, install the thin lock nut on top of each of the connecting rod nuts to prevent them from loosening.

22 Bolt on the cam gear, and it's ready to install.

23 Regardless of whether you are installing a new or old camshaft, the cam gear has to be installed. There is a slotted key on the end of the camshaft that matches the slot on the cam gear.

24 You can see the two bolt holes that secure the camshaft. The other three bolts attach the front mounting plate to the engine. The front mounting plate has to be installed before the cam is installed. After the camshaft bearing is installed in the block, you're ready to slide the cam in place. Once the cam is installed in the block, two bolts hold it in place. You can reach the two bolt heads by turning the cam gear.

25 *There is a timing mark on the crankshaft and the camshaft. The timing mark on the crankshaft and the timing mark on the camshaft must line up. If not, the engine will not start or run. It is hard to fix this once the engine is installed in the Jeep. If you haven't lined up the timing marks before the engine is installed, you will have to remove the radiator, the front timing cover, and the camshaft gear. Then, you can align the timing marks and reinstall everything. (If you are forced to do this, get an extra camshaft gear because it may not come off easily.) Note that Willys MBs, Ford GPWs, and early CJ-2As used a timing chain instead of timing gears on L-head engines.*

26 *Replace the timing cover and seal. This seal is in the timing cover, and the crankshaft pulley slides onto it. If you have a leak from the front of the engine, it's the timing cover. To change the seal, remove the radiator and the fan blades. Use a gear or wheel puller to remove the lower pulley. Then, remove the timing cover and change the seal. It's very important to install it straight the first time.*

27 *The side cover is on, and the engine is almost back together. With the head gasket installed, it looks great, but lots of work went into this.*

28 *From each angle, you can see the completed 4-cylinder flat-head engine. An F-head engine would have gone through the same procedure. Two freeze plugs are visible. Most engines have them, and most machine shops replace them. There are designed to pop out of the block if the coolant freezes. This allows the frozen coolant to expand and not crack the block. Replace them during any engine rebuild because they can become rusty and leak.*

Take the engine to the shop. Let them take it apart, clean it, and inspect it for cracks. Let them do the machining on the block, replace the valve guides and cam bearings. The shop can bore the cylinders and turn the crankshaft.

It can give you the specs on the rebuilt block. The shop will tell you how many thousandths it has bored the cylinders and how many thousandths it has turned down the crankshaft, etc. You can order an engine rebuilding kit with everything you need to put it back together. The kit comes with everything: new pistons, rings, bearings, gaskets, seals, etc. The machine shop does the machining, and you put it back together. You will save money on the labor and have the satisfaction of doing it yourself. There are a number of good service manuals that can guide you through the process. Keep in mind that the machine shop will not guarantee *your* work.

Letting the Machine Shop Do Everything

Remove all accessories before delivering the block to the machine shop. That includes removing the distributor, ignition coil, water pump, (I leave the oil pump and have them rebuild it for me), both the intake and exhaust manifolds, oil filter, sending units for both oil and coolant, and the thermostat housing. The block should be bare, and that is the way the shop wants it. If you leave any accessories on the block, you're just going to get it back in a spare parts box. The machine shop's only job is to machine and reassemble the block.

Use tape and a marker to number the spark plug wires with their cylinder number before removing them from the spark plugs. Leave the spark plug wires attached to the distributor cap until you are ready to replace them. If you are installing new spark plug wires, match the lengths to the old ones on the distributor cap and only replace one wire at a time. Move the tape from the old wires to the new ones so that you keep the firing order correct.

If you had any broken bolts on your block or they were rusty and you broke them taking it apart, don't worry. Show the shop each one, and it will drill and tap the holes so you can put it back together with new bolts. The most common ones to break are manifold bolts, head bolts, water pump bolts, thermostat housing bolts, etc. Two small studs hold the ignition coil to the block. Some may have rusted or broken off. The hole in the block is threaded, and many people just replaced the threaded stud with a bolt. The same can be done for the exhaust manifold bolts.

Some blocks have studs and some have bolts. Both work. Note that the manifold bolts in the L-head and F-head engines have both coarse and fine threads. Match up the proper nuts.

Some machine shops assemble all the accessories for you, but they will often charge you a lot. Besides, if they give you back the engine ready to install in the Jeep, what are you doing? After they disassemble the block, they will steam clean it or boil it to remove all the grease and dirt.

Once the block is clean, the machine shop will inspect it for damage and cracks. Keep in mind that most old Jeeps have engines that are 30 to 60 years old. Some have been rebuilt several times. If they have a small crack, you have a choice: the shop can weld it and it will be okay, or you can find another block. Be aware that finding uncracked L-head and F-head blocks can be hard to do.

The machine shop will also let you know how large the cylinder bore is. The size of the bore and the piston is stamped on top of the piston. It will say 20, 30, etc. (up to 80). I do not like to bore out a block to more than 60 thousandths. By 80 thousandths, the cylinder walls are getting too thin. However, you do not have to get another block if that is the case. The shop can put sleeves in the cylinders and bring them back to standard size.

The machine shop keeps track of all the work it does to your engine. Most provide a computer printout of the parts and labor. Do not cut corners on the rebuild. The shop already has it apart.

Let the shop resurface the face of the flywheel. That takes out all the marks and grooves and gives the clutch a better surface to grip. Let the shop resurface the exhaust and intake manifold (sandblast or bead blast). This gives a nicer look to the finished product. Also have the shop remove all the head bolts and studs and surface the block and head.

Most importantly, when these engines were first manufactured, we all used leaded gas. Make sure the shop installs new valve guides that are compatible with unleaded gas.

If the shop does not ask you about painting the rebuilt engine, ask it to paint it the color you want. L-head and F-head engines were painted gloss black. Buick Dauntless V-6s used a GM blue/green engine paint. AMC 4-, 6-, and 8-cylinder engines were painted metallic blue. Otherwise, you can add your own style by painting the engine any color you want. Heat-resistant engine paint is available at most auto parts stores.

Cooling

Now that your engine has been rebuilt, you must keep it cool. There are a variety of parts working together that are tasked with accomplishing this.

Radiator

A radiator is the most important part of the cooling system. While the radiator is out and your Jeep is waiting for the engine and body to be finished, take the radiator to a radiator repair shop. If your Jeep has a heater, also take the heater core with you. Ask the radiator repair shop to pressure test the radiator and the heater core. It will check for leaks. Some radiators may be original, and they can become clogged and leak.

Radiators have a central core with vertical channels. Attached to them are rows of fins that increase the surface area and improve the cooling effect. An upper and lower tank hold coolant, and the water pump on the engine pumps the coolant through hoses to the radiator where it is cooled and returned to the engine. The mechanical fan blades mounted to the water pump provide cooling by pulling cooler air from the outside through the grille and passing it through the radiator to cool the hot coolant.

A metal mounting flange is soldered to the outside of the radiator and has four to six holes in it that are used to bolt the radiator to the inside of the grille. Make sure you ask the shop to check the welds from the bracket to the radiator.

Also, on the bottom of the radiator, there is a drain valve. Make sure it works and you can close it. If not, replace it. Try to replace it before you go to the radiator repair shop in case it breaks off.

Thermostat

The thermostat regulates the temperature in the engine. Remember that water boils at 212°F. You do not want the coolant in your engine to get that hot. It can damage the gaskets and warp the heads.

When the engine is cold and you start it, the thermostat is closed and stops the water in the engine from flowing. This heats the coolant and allows the engine to reach normal operating temperature.

At a preset temperature, the thermostat opens and allows the coolant to flow into the radiator and get cooled. The rating on your thermostat depends on your location. It's a matter of warm climates versus cold climates. Most thermostats start at about 175°F and go to 195°F. The colder the climate, the hotter you want it to run. You can hear the thermostat open and the coolant start flowing.

Most people like to keep their Jeep original and use the electronic dash gauges that come with the Jeep. I like to use the mechanical gauges that show the temperature, and you can watch the arrow move from 200 to 185 when it opens. You never want to see an engine overheat.

Keep this in mind: thermostats are cheap. Sometimes they are defective and do not work, even when new. Keep an extra one or two around when installing a new engine. Once they work, they last for a long time.

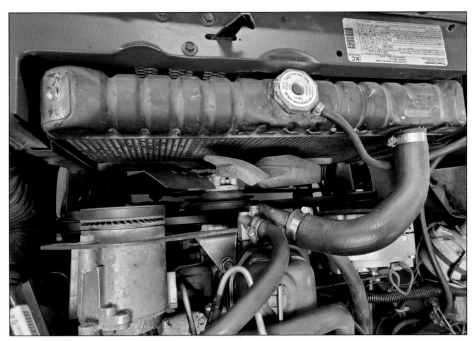

Inspect your cooling system before you reinstall it. The radiator should be flushed and pressure tested before using it again. Rubber hoses are easy to inspect before installing them, and they are fairly affordable to replace. The hose clamps should be replaced at the same time. The water pump may not be leaking now, but it is easier to replace before the fan and belts are in place. Changing the thermostat and gasket is easy, but only use quality parts. Some thermostats may be defective from the store. The same is true for radiator caps. Then, just add a fresh 50-50 mixture of traditional coolant.

Hoses

Radiator and heater hoses deteriorate over time. They should be firm but springy. If they are crunchy hard or spongy, replace them. New hoses and clamps are cheap insurance against being stranded somewhere with a leaking engine.

Water Pumps

If the engine has been sitting for years before being rebuilt, the water pump has had stagnant coolant resting inside, and the rubber parts are rotten. This can cause a leak later on. Although replacing a water pump later is fairly easy, you may as well mount a fresh one at this point. They are very simple and affordable.

Dressing the Engine

Once the engine is ready, start to attach the accessories. Check to see that there are no bolts or studs broken off on the block. These include the manifold studs, ignition coil studs, etc.

Now is the time to use your thread restoration tools to clean out all the threaded holes in the block. For example, use them on the water pump, fuel pump, manifold, thermostat housing, and oil and temperature sending units.

Get your gaskets lined up and make sure you have gasket sealer. Try using Permatex; it's sloppy but it works.

It's a good idea to also have a roll of Teflon tape. You'll need to wrap this on any threaded parts, such as the water sending unit and the heater fitting that goes into the water pump and into the head.

You can install the accessories in any order. Some people like to install everything associated with a particular system, such as the cooling system, the fuel system, the electrical system, the exhaust system, the intake system, etc.

Add the sheet metal, including the oil pan, timing cover, and valve cover(s) when it is convenient. Wait to install the fan. Each Jeep model has a different-size radiator shroud. The fan and the shroud may get in each other's way, so wait until later.

Do not put the motor mounts on the engine until the engine is bolted to the bellhousing. Otherwise, they will be in the way when you are maneuvering the engine to get the bellhousing bolts in.

Installation of the generator or alternator can wait. They may be in the way when you install the engine.

Block Components Assembly

1 *Using the clutch pilot tool, place the clutch disc and the pressure plate on the flywheel. Hold them in place while you line up and thread in the mounting bolts. Thread the bolts finger tight but loose enough that you can align them with the clutch pilot tool.*

2 *Center the pressure plate using the pilot tool and use a torque wrench to tighten the mounting bolts in a star pattern. Remove the pilot tool from time to time to make sure everything is still lined up and centered. Continue tightening the bolts until you reach the proper torque settings described in the service manual or the clutch kit instructions.*

3 This is a freshly machined flywheel ready to accept the clutch components. You can see the six threaded holes ready to accept the pressure plate. Note that you can no longer use an engine stand at this point. Installing the clutch requires access to the flywheel.

4 The carburetor is easy to mount on an F-head engine. But before you do it, check the head bolt torque near the manifold. Once the carburetor is installed, you won't have access to it. Once the head bolt is tight, lay down the carburetor gasket and sit the carburetor on top of it. Then, bolt it down. F-head engines often have a tubular lifting bracket on top of the engine. It is optional; you can keep it or remove it. The black bracket bolted to the valve cover holds the spark plug wires. They are often missing, but reproductions are available.

5 Both the L-head and the F-head engines have a simple exhaust. The L-head has a two-piece exhaust manifold, and the F-head engine has a one-piece manifold. The L-head manifold has a heat riser in the lower half. This heat riser valve restricts the exhaust and causes heat to warm the manifold. You may need to apply the manual choke during this process to get the carburetor to function well. The heat riser valve rotates on a counter-weighted shaft dampened by a bimetallic spring. As the spring relaxes, exhaust gas pressure can force the heat riser valve to open more easily. In some manifolds, the heat riser valve is rusted in place. Check to verify that yours is working properly. Repair kits are available.

6 *There are many types of oil filters but only a few brackets. On this F-head engine, you see a standard oil filter canister with the correct bracket.*

7 *The oil filter bracket on the left is for an L-head engine, the bracket in the middle is for an F-head engine, and the one on the right is an L-shaped bracket that someone made and mounted on a fender or firewall.*

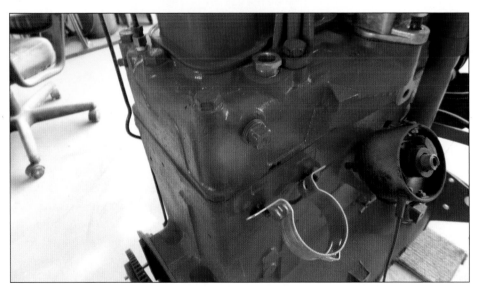

8 *The ignition coil bracket slides over the two 1/4-inch studs and is bolted tight with the ignition coil inside. You may find that the studs have broken off in the past. They can be drilled out, tapped with new threads, and bolts can be used instead.*

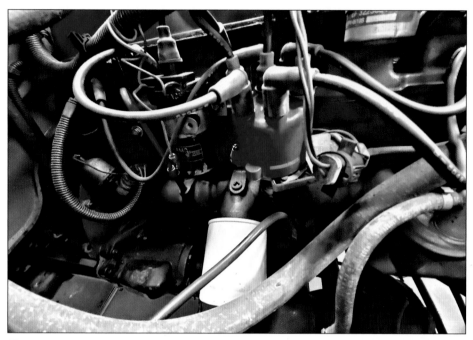

The ignition system on an older Jeep requires more frequent maintenance than on a new vehicle. Performing a simple tune-up helps the engine start and run smoothly. Replacing the spark plugs and distributor points is a good start. Be sure to set the gap according to the service manual. Most tune-up kits also include a new rotor, distributor cap, and condenser. Consider replacing the spark plug wires if they are old.

Starting a New or Rebuilt Engine

First, add oil and coolant. Be sure to use a break-in oil. Talk to the machine shop that rebuilt your engine and get their advice on the type of oil to use. For the radiator, use at least a 50-50 coolant depending on your climate. Check the radiator/heater hose clamps and belts for tightness. Make sure the battery is secure and the cable ends are tight.

It may be a good idea to perform a tune-up at this point. Replacing the spark plugs, points, and condenser is pretty easy. Use a feeler gauge to adjust the gap on the spark plugs and points. Your service manual provides the correct gap. Using a timing light and dwell meter allows you to fine-tune the engine's performance.

If the carburetor was simply removed and reinstalled, it should be

Installing the Drivetrain

It is much easier to install the accessories while the engine is on an engine stand than it is to install them onto the engine after is installed. Now, it's time to put the engine back in.

If the body is off, put the entire drivetrain together and place it onto the chassis. Bolt the transmission crossmember onto the chassis. Then bolt the engine, transmission, and transfer case together. Next, lower the engine/transmission/transfer case onto the chassis landing on the crossmember. Now you have great access to the motor mounts and both of the transmission mounts.

Then, install the driveshafts, speedometer cable, and parking brake cable, and you're ready to install the body.

If nothing has changed between the old engine and the rebuilt engine, you may be able to set the carburetor back on the manifold and reattach the throttle linkage, vacuum hoses, etc. Once the Jeep is running again, you may want to adjust the idle speed and/or fuel mixture. Consulting the service manual will help you do this.

fine. Idle speed and air mixture can be adjusted with a screwdriver. Again, your service manual provides the details for your specific carburetor.

Once the engine has been run, check and refill the coolant level. Any air bubbles should have escaped. Fill or top off the gear oil in the transmission, transfer case, and axles before driving the Jeep. Of course, automatic transmissions need to be filled with the correct automatic transmission fluid (ATF), and the fluid level should be checked after a short drive.

Oiling the Engine

Oil is one of the most important but often overlooked elements in running your Jeep. It is the heart's blood to your engine. Replace the break-in oil after 500 miles or so. It is not recommended to use synthetic oil in the early 4- and 6-cylinder engines. A simple SAE 30-weight oil works best. Change the oil and filter every 3,000 miles, and your engine will last a long time. Most Jeep engines will take 4 to 5 quarts of oil. Refer to your service manual for your specific engine. It may vary when changing the oil filter at the same time. Check the dipstick repeatedly so that you do not overfill the engine.

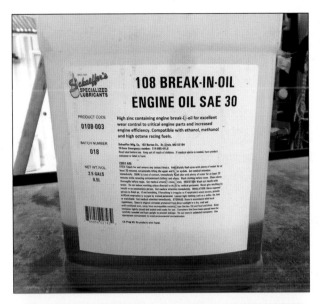

Let's start at the beginning with a new or rebuilt engine. I like to use Schaeffer's 108 break-in oil in SAE 30 weight. It comes in 2½-gallon cans. It was recommended by my machine shop and is available online and through automotive speed shops. Use it for 400 to 500 miles and then drain it out. Replace it with traditional SAE 30-weight oil. Do not use synthetic oil!

Oil-bath air filters were used from 1941 through 1971 in World War II Jeeps and civilian CJs with the L-head and F-head engine. A small tray of oil in the bottom traps dust and dirt when air flows through the filter. It's common to change the oil in the air filter during each engine oil change.

This is a good look at the inside of the oil-bath air filter. Simply pour oil into the bottom tray until it reaches the full line indicated on the outside of the tray.

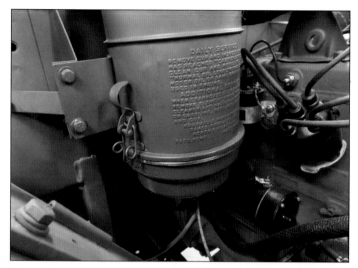

Even the early MBs and GPWs had oil-bath air filters. Two clamps on the outside of the air filter release the bottom tray when loosened. All Jeeps come with an engine oil filter. Two basic sizes traditionally came on 4-cylinder engines, whether F-head or L-head. They have drop-in cartridge filter elements.

Exhaust System

From 1941 to 1945, the MB and GPW had a simple header pipe and a flex pipe that curved around under the transmission/transfer case skid plate. It came out under the passenger seat where the muffler was attached with a short tailpipe exiting in front of the rear passenger tire.

From 1945 to the 1980s, not much changed; there were only two types of exhaust systems. The civilian version used a conventional automotive exhaust system with hangers and clamps. The M38 and M38-A1 used the same muffler as the MB and GPW, but with a

Replacing the exhaust is simple. Start by placing the hangers in the appropriate locations for your Jeep. You can install the header pipe, but do not tighten it down until the engine is in place. Now, you can attach the header pipe to the straight pipe. Once the straight pipe is in place, add the muffler. Finally, you can add the tailpipe.

Make sure the tailpipe does not touch the chassis, springs, or rear bumper. If it does, it vibrates and rattles. Do not tighten any clamps or brackets until you have the whole exhaust system in the correct position.

triangular flange at each end that used a gasket and three bolts to hold the system together, making it almost waterproof.

If the body is off and you're getting the chassis ready for the engine and the body, you can install the exhaust system. Most of it bolts to the chassis, and only a few brackets bolt to the body. Put on as much as you can now while it is open to you. Once the body is on, you have to work from underneath. Don't overtighten the brackets or hangers. You may have to adjust the muffler and brackets later.

The MB and GPW are the only Jeeps where the tailpipe comes out in front of the rear tire. All of the other Jeeps have exhaust systems that exit at the rear.

In the MB and GPW, the header pipe makes a 90-degree turn behind the bellhousing, crossing over to the passenger's side on top of the skid plate for protection. It then makes another turn to attach to the muffler.

TRANSMISSION AND CLUTCH

For the first 30 years, the Jeep used basically the same transmission. From 1941 through 1945, it used the T84 transmission. Remember, it was 1940 when the U.S. government put out the request for proposals for a new Army 4x4 utility truck to carry two to four men and supplies over rough terrain.

The technology before World War II was not as advanced as it was after the war. Also, keep in mind that in the auto industry it takes two to three years to research, design, and build a new car or truck. However, three auto manufacturers had 4x4 prototypes ready in just one year.

In 1945, when World War II was winding down, Willys switched to the agricultural market for its production model CJ-2As, so some changes and upgrades were made to the Jeeps for farm use. One thing that was improved was the transmission. The upgraded transmission was then used for another 25 years in the CJ-2A, CJ-3A, CJ-3B, and early CJ-5s. It was called the BorgWarner T-90. It was a larger and stronger version of the 3-speed manual T-84 and bolted up to the same Dana 18 transfer case.

The T-84 has a main shaft going through the center of the cluster gear. The T-90 has the same design but with 88 needle bearings in it. The larger T-90 case gave it room for larger front and rear bearings and allowed faster speeds.

In late 1945 and 1946, the CJ-2A had a 3-speed column shifter. The same T-90 case and gears were used with a side shifter (column shift linkage) instead of the top shifter. In 1947, it switched back to the top shifter. This is one example of how versatile the T-90 could be.

Side to Top Shifter Conversion

The column-shift T90 can be easily converted from column shift (transmission-side shift) to floor shift (top shift) by installing a floor-shift top cover assembly. Take the internal components out of the side-shift version, remove the side forks, and plug the two linkage holes.

In 1970, American Motors Corporation (AMC) purchased Jeep. After 25 years of the T-90 (1946–1971), AMC decided to use four new transmissions. The one being rebuilt and featured here is a T-90.

Most 3-speed standard transmissions work the same. By this time, you should have a service manual for the model of Jeep you're restoring. By reading this manual, you can rebuild the transmission.

Shifter Assembly

The only repair you can do on the transmission while it is still in the Jeep is to change the shifter assembly. If you're in first gear and you have to lean way over almost to the dashboard to get to second gear, there may be too much play in the shifter assembly.

Other than changing the shifter assembly, all other repairs to the transmission need to be done with the transmission out of the Jeep.

Removing the Floor Covers

The MB, GPW, CJ-2A, CJ-3A, and CJ-3B models had floor covers that only exposed the shifter.

In 1950, when the M38 came to market, Willys made updates to the floorpan. In the M38, M38-A1, CJ-5, CJ-6, and CJ-7, all of the floor covers come out to expose everything from the bellhousing bolts to the transfer case. This was a huge improvement for access.

Removing the Transmission

After you remove the floor cover and shifter assembly, place a rag into the top of the transmission so that foreign objects and dirt will not get into the transmission. If you have a hydraulic car lift, the transmission and transmission case can come out as one unit. Be aware that these two components are very heavy. The following steps are needed to remove these two items.

1. Remove the floor covers.
2. Disconnect or remove the front and rear driveshafts.
3. Disconnect the speedometer cable and parking brake cable.
4. The 1941–1946 Ford GPW, Willys MB, and CJ-2A have an engine stay cable that connects the bellhousing to the crossmember. Remove it.
5. Disconnect the clutch pedal spring from the clutch bellcrank.
6. Remove the four bolts holding the bellhousing to the transmission or disconnect the bellhousing bolts to the engine leaving the bellhousing on the transmission.

Next, unbolt the transmission crossmember from the chassis. If you are using a hydraulic car lift, lower the Jeep close to the floor and put blocks or a transmission jack under the transmission. Unbolt the crossmember and then raise the Jeep, leaving the transmission on the floor.

You can also use an overhead chain hoist. Wrap a chain around the transmission and hook it to the overhead chain hoist. Unbolt the crossmember and lower the transmission to the ground. Then, hook the chain hoist to the Jeep and lift the Jeep. After that, slide the transmission out from underneath the Jeep.

If you removed the transmission and transfer case together, separate them before work can begin on either one of them.

Disassembling the Transmission

1 *Before you start rebuilding the transmission and transfer case, make sure you have both gasket sets. There may be a little overlap by having a few duplicates, but it's better to have a few extra than to be missing one. Most auto parts stores have gasket paper in different thicknesses if you would like to create your own gaskets. Simply trace the old gasket onto the new gasket paper and cut it out.*

2 Most Jeep aftermarket parts suppliers have rebuild kits for the T-90 and other Jeep transmissions. Find your specific kit and order it.

3 Pressure wash the transmission and transfer case. Do not worry about getting water inside of the case. You can drain it, and when the case is empty, dry it with a towel or rag. For grime that won't blast off, use a brush and a good detergent.

4 Drain all the gear oil out of the transmission and the transfer case. Open the top of the transmission by unbolting the shifter cover. At this point, you will be able to see any chipped or broken teeth on the gears.

5 Remove the five bolts on the round cover on the back of the transfer case. Look inside and you will see the nut that holds the transmission to the transfer case. Remove the cotter pin and nut and then reach in with your hand and grab hold of the gear. If you turn it left and right a few times and pull it back, it should come out. Now, you're ready to separate the transmission from the transfer case. Five bolts hold them together. Unbolt them, and the two units come apart.

6 Remove the 10 bolts holding the transfer case cover onto the transfer case.

7 You will be able to see all the gears once the cover is removed. If all the teeth are in good condition and there is no play in the bearing, leave it alone. If you have not driven the Jeep and you want to be sure everything will be in proper, working order, remove the gears and replace everything.

8 *Use an Allen wrench to remove the front collar. There are three bolts that have Allen heads. Bag and tag them for the parts washer once they're removed.*

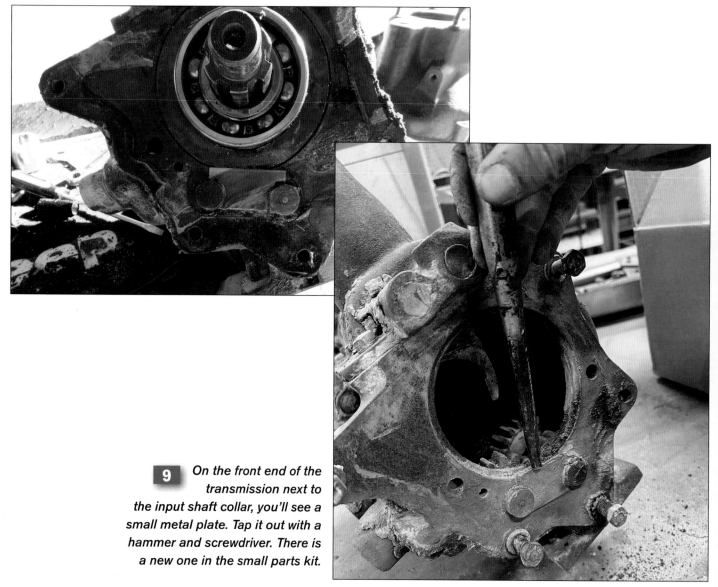

9 *On the front end of the transmission next to the input shaft collar, you'll see a small metal plate. Tap it out with a hammer and screwdriver. There is a new one in the small parts kit.*

10 *Sometimes the front bearing needs an assist to remove it. A small wooden 2x4-inch block and a hammer helps.*

11 *The top shaft in the transmission should come out easily. However, you're not ready to put in the new parts yet. The cluster gear on the bottom has to come out.*

12 *Around the reverse idler gear is an oil splash. Use the Allen wrench again and save the bolts.*

13 *The purpose of the oil slinger is to splash oil from the bottom of the transmission to the top. The top of this one is a little chewed up from the last time someone took it apart and didn't know the correct procedures.*

14 *Take it all the way out. If you do not, you can't remove the cluster gear.*

15 *Use a rod that is a little smaller than the cluster gear shaft and use a hammer to drive it through the transmission, pushing out the metal shaft from the cluster gear.*

16 *Once the shaft is out, lift out the cluster gear.*

17 *Take another look at your main gear shaft for chipped teeth. Replace the synchronizer assembly.*

18 *To change the synchronizer assembly, remove the C-clip (some people call them snap rings). Once the clip is off, the synchronizer assembly slides off the shaft. Put the new one on and replace the clip.*

19 *Now, thoroughly clean out the case. You might see small metal shavings from wear and tear of the transmission from years of work. If you see large brass or metal pieces, there was something wrong in that transmission at one point. Thoroughly examine the internal components.*

A front transmission seal leak can easily be mistaken for a rear main engine seal leak.

Some talented machine shops and good mechanics can change the rear main engine seal without removing the engine. If the rear main seal is replaced and you still have a leak in this area, it is most likely the front transmission seal. This can only be changed with the transmission out of the Jeep.

Just remove the front bearing retainer cap. Then, remove the three bolts that have Allen heads. The retainer cap will slide off the input shaft. Replace the felt seal and replace the gasket, making sure you use a good gasket sealer, such as Permatex. Then, replace the bearing retainer cap.

It is much easier to fix the transmission once it is out. If you have to change the synchros, gears, or bearings, replace the front seal last after everything has been replaced.

Reassembly of the T-90 Transmission

1 *Always have a good supply of grease on hand. Use it to pack the 14 needle bearings into the open end of the input shaft. If you do not pack them in with grease, they fall out and you will never be able to put the input shaft back in the transmission.*

2 These are a few of the home-made tools you'll need to reassemble the transmission. Use a long rod or wooden dowel to drive out the metal shaft in the cluster gear. Then use a 5/8-inch wooden dowel to install the 88 needle bearings into the cluster gear. Use the same 5/8-inch wooden dowel to load the needle bearings. Finally, use a short piece of 3/4-inch conduit that slips over the wooden dowel to allow you to insert the bearings into the cluster gear.

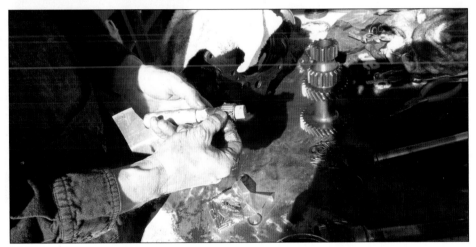

3 First, put one of the spacers on the shaft. Next, grease 22 needle bearings and wrap them around the shaft. Then, place another spacer, another 22 needle bearings, and another spacer.

4 Here are two sets of 22 needle bearings with spacers ready to go into the gear.

5 *Slide them down as far as you can without forcing them.*

6 *Using the conduit, slide them in.*

7 *There should be a sleeve in the middle of the cluster gear. It's about 1½ inches long. You will wind up with the sleeve in the middle, two sets on one side, and two sets on the other side so that the ends look like this.*

8 *Clean the main cluster gear shaft. Wipe it down with a small piece of emery paper.*

9 *Put a little grease or oil on the shaft so that it goes in easily.*

10 *Using your main shaft rod, push the wooden dowel out of the cluster gear. Watch the groove in the end of the metal shaft and make sure it faces the other one from the reverse idler gear. On the table is a large washer. There are three of them and they go on either end of the cluster gear. That fills up the space so that the cluster gear does not move left or right in the case. (Note: if you can only get one on each end of the cluster gear, that is okay.)*

11 *Slide the new key in place. Tap it back in with a hammer and screwdriver.*

12 *It is okay to have the key overlap on the main shaft plate.*

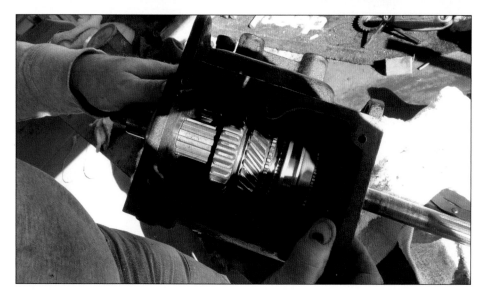

13 *The top and bottom shafts are in place.*

14 Sometimes, you may need to tap in the front bearing.

15 Put a new seal on the front input shaft. The felt ones are very good, but seals made from other materials are available also.

16 Install the new gasket and the input shaft cone.

17 *Tighten the three Allen-head bolts. Some people lose them and use a regular bolt. If the bolt head is too close to the edge, the bellhousing may not fit.*

18 *Install the shifter, making sure the two forks are in the correct slots on the synchro and the gear.*

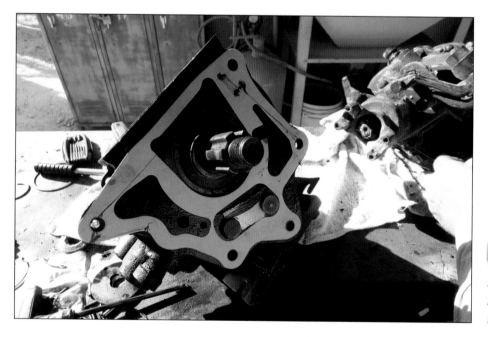

19 *Use Permatex Gasket Shellac Compound and cover both sides of the gasket. This gasket fits between the transmission and transfer case.*

If you have to replace any gears on the top transmission shaft, it should come apart in two pieces.

At that point, you can change the synchros, slider gear, and/or the front/rear bearings. If any gear teeth are broken or missing on the bottom cluster gear, you will have to remove the bottom cluster of the transmission.

After you have done the repairs, reverse the process, and put it all back together.

Always remember to add the fluids back in before driving the Jeep. An 80W-90, API-GL5, or MT-1 rated fluid is a good choice.

The transmission will have a lower drain plug and an upper fill plug. Remove the upper fill plug and use a gear oil pump attached to a bottle of gear oil. Pump the fluid into the upper fill plug hole until it begins to dribble out and overflow. Once the oil starts flowing down the side of the transmission, stop pumping and reinstall the fill plug.

DANA 18 TRANSFER CASE

Come to a Complete Stop

Whenever you are using either one of the transfer-case levers (2WD/4WD or high/low range), always bring the Jeep to a stop before moving the levers. ■

Jeep used the same Dana 18 transfer case for 30 years (1941–1971), so it is a commonly found item. Two levers come through the floor to operate this transfer case. In the early Jeeps from 1941 until 1949, both levers were the same length. After that time, the one on the left was made longer. The lever on the left is used to select 2WD or 4WD. You can engage 4WD by pulling back on the lever.

The lever on the right has three positions. Shifting to the center position puts the transfer case in neutral. No matter what transmission gear you select, the driveshafts will not turn. This position is selected when you are using a PTO or flat towing. When you pull this lever back toward the seat, it goes into high range. This is for driving at normal speeds in both 2WD and 4WD. When you move the lever forward, it shifts the Jeep into low range. In low range, the Jeep will not travel more than 5 to 10 mph. This position increases the torque and is used for towing heavy objects, snow plowing, or pulling another car or truck out of the snow or mud.

The Dana 18 transfer case was used in more than 1 million vintage Jeeps. They are very common, and parts are readily found. Luckily, they are also simple to rebuild. The two levers on the right operate the transfer case. One lever engages 4WD and the other engages the high-range or low-range gears.

Front and Rear Seals

There is a front and a rear seal on the transfer case where the driveshafts attach. They commonly leak, but it is not too difficult to replace them.

Start by removing the four bolts holding the skid plate to the transmission. While the skid plate is off, give it a thorough cleaning and

Replacing the front or rear output yoke seals on a transfer case is common due to leaks. Removing the old leaking seal is the first step. Then, slide the new seal over the splined pinion shaft and align it so it stays in place.

Use a 1½-inch socket and begin tapping gently. The new seal must go in straight, otherwise it will leak.

repaint it. The crossmember supports the transmission while you work.

If you have the use of a hydraulic car lift, that is best. If not, you can do this repair on a level spot of ground. If working on the ground, place the

Only drive the new seal into the housing until it's flush. Driving it any farther will cause a new leak.

Jeep on four jack stands and remove the wheels. This provides some room to maneuver. Once the Jeep is securely raised, remove the front and rear driveshafts. If only one is leaking, just remove the one you have

to replace. However, it can be a good idea to replace them both at the same time while you're under the Jeep.

Once the driveshafts are off, look at the yoke on the front of the transfer case. There is a large nut in the middle of the yoke. Remove the cotter pin first and then the large nut. This is easier with an impact wrench. If you do not have an impact (air) wrench and you are using hand tools, you might need a large pipe wrench to prevent the yoke from turning while you remove the nut.

Once the nut is off with the washer behind it, tap the yoke lightly with a hammer from the back and it will come off. This exposes the leaking seal.

Seal Pullers

Every seal puller that I have purchased has broken before I removed the first seal. So, I use a cold chisel. With a hammer and chisel, start at 12 o'clock, and every 10 degrees or so, cut a notch in the old seal. It is brass and rubber, so it cuts easily. Once you have an abundance of cuts in it, use a large screwdriver or pry bar to pry it out. The housing around the seal is part of the transfer case and is steel. You will not damage it if you miss once or twice.

Be careful not to hit the housing or the threads on the shaft that holds the yoke. Some people put the nut back on to protect the threads. Once the old seal is out, use a small piece of emery paper to clean the edge of the seal housing, making sure there are no burrs on the edge. Then, wipe a finger-full of oil or grease on the inside of the round opening where the new seal will sit. (It is a good idea to order a few extra seals, just in case one is damaged.)

Put the new seal in place. It will not go in on its own. You will have to press it in. I use a 1½-inch deep socket. Place the socket on the seal, and the pinion shaft will go through the center of the socket. With the socket resting on the seal, evenly tap the socket with a hammer to make sure the seal goes in evenly.

Only drive the seal into the housing until it is flush with the opening. It is deeper than it needs to be, and it is easy to drive the seal in all the way and cause a leak. Do not do that. Only drive the seal in until it is flush with the edge of the housing. Once you have the new seal in place, reinstall the yoke, washer, and nut. Then, replace the driveshaft, and you are done.

Now, on to the rear seal. This one is a lot more involved, but this is not as big of a deal as it looks. You just have to remove the parking brake assembly to get to the seal. Take off the four bolts that hold the rear driveshaft to the back of the transfer case. They should be in the center of the parking brake. Once the rear driveshaft is off, remove the parking brake.

Disconnect the parking brake cable or floor lever rod on the M38 and M38-A1. Depending on the model, there may or may not be more bolts to remove the parking brake. Check the service manual. Once the parking brake is removed, use the same process for the replacement of the front seal. It is the same seal with the same part number.

After the seal is replaced, reverse the process and replace the parking brake and rear driveshaft.

Driveshaft(s)

With the driveshafts (propeller shafts) removed, it is a good time to clean and paint them. It's also a good time to have the U-joints replaced and have the driveshafts balanced to prevent vibration.

You should be able to find driveline service shops in your area, but they can be shipped if needed. Once you reinstall the driveshafts, grease the U-joints and any other grease fittings that are present with a red tacky automotive grease.

Case Mounts and Bearings

There are two mounts. The one under the transmission is a long rectangular mount, which is held onto the transmission with two bolts, and it also attaches to the crossmember with two bolts. The mount is easy to change when the transmission is out of the Jeep.

When you inspect your Jeep and find it has to be replaced, this is what is recommended:

Using a hydraulic car lift is the easiest, but you can work while lying on the ground.

Use a floor jack to support the transmission. Don't try to lift it; just hold it in place. Then, unbolt the crossmember and lower it. The transmission will be supported by the jack while you work.

Now you can replace the transmission mount. It works best if you bolt the new transmission mount to the bottom of the transmission first and then reinstall the crossmember and bolt the mount to it.

While you have the crossmember out, replace the transfer case mount if needed. It is inches away on the crossmember and only has one large bolt. Always make sure that both mounts are in good condition. With exposure to the elements, heat, and stress, the rubber will break down.

While the crossmember is out, remove the 10 bolts holding the lower inspection cover on the transfer case, drain the fluid, and drop the cover. You will have a great view of the gears. As long as they don't have any chips on the teeth, in most cases they are okay.

The only other things that can go bad on the transfer case are the front and rear main bearings. You'll know when that happens. The Jeep will sound like an old school bus when you are driving. That noise could also be the main bearings in the transmission. If you experience any noises like that, change all of the bearings. To change them, the transfer case must be removed from the Jeep.

It is a good habit to check the fluid levels regularly in the engine, transmission, and transfer case. Most of the damage done to a transmission or transfer case is caused by low fluid levels. They overheat and burn up over time.

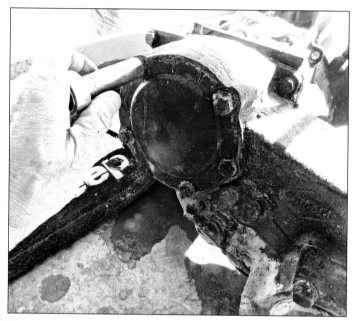

To separate the transfer case from the transmission, remove the five bolts that fasten the transmission and transfer case housings together. Then, remove the five bolts from the round cover plate.

Next, remove the large nut and cotter pin to separate the transmission and transfer case. Try not to drop the cotter pin in the transfer case.

Transfer Case Removal

1. Remove the front and rear drive-shafts. Remove the parking brake assembly.
2. Disconnect the parking brake cable and the speedometer cable.
3. Drain the fluid.
4. Remove the round cover on the back of the transfer case directly behind the transmission. It has five bolts.
5. Look inside. You will see a large nut. Remove the cotter pin and that nut. Once they are off, reach in with your hand and pull the large gear in the center back toward you. Turn it a little, and it should come off with no problem.
6. Unbolt the round transfer case mount. There is only one; it is the round one with the large nut.
7. Inside the Jeep, take the shift lever knobs off the shift levers and remove the rubber boot on the floor. This allows the transfer case to drop and be removed.
8. When all of this is complete, there are only five bolts holding the transfer case onto the transmission. Remove them and carefully wiggle the transfer case until it separates from the transmission. Be careful; the transfer case is heavy. If working on the ground, use a floor jack to support the weight. If working on a lift, use a transmission jack or a helper.

You can remove the transfer case at any time without worrying about the engine or transmission falling. They are held in with their own mounts.

Because the same transfer case was used for 30 years in many Jeep models, it is possible to replace yours with a newer or older one in good condition. The shift levers can be changed to match the Jeep model on which you're working.

If you have a 1941–1949 MB, GPW, or CJ-2A with the two short levers, you can use a newer transfer case but retain the short levers.

If you have a CJ-3A, M38, CJ-3B, M38-A1, or CJ-5, you can reuse the longer levers from an older Dana 18 transfer case.

The Dana 18 transfer case was eventually replaced by the Dana 20 transfer case found in the 1972–1979 CJ-5, CJ-6, and CJ-7.

The BorgWarner 1339 Quadra-Trac transfer case was used in 1976–1979 CJ-7s with the TH400 automatic transmission.

Finally, the Dana 300 transfer case was found in all 1980–1986 CJ-5s, CJ-7s, and CJ-8s. Be aware that the 1980 version had a shorter 3½-inch rear output housing assembly. The 1981–1986 version has a 5½-inch rear output housing.

BODY AND INTERIOR

Jeeps receive their share of dings and dents when driven off-road, and they tend to rust if left sitting outside. So, some amount of body repair is usually necessary. If the labor and materials required to fix the original body are excessive, it may make more sense to replace the body with a reproduction. Either way, the body will need to be painted.

If your goal is to learn every step in the restoration process, you may want to take a class at a nearby vocational school to learn the basics. Jeeps are pretty easy to work on and are forgiving of mistakes.

If you are looking for a fun trail Jeep, your body work skills may be fine. If you are looking for a museum restoration or one that will be judged in a show, you may want to pay a professional.

Either way, having a factory service manual is essential to answer questions about how things are supposed to come apart and go back together. They are available as reprints.

Body Condition Analysis

The factors below should be considered when analyzing a Jeep body's condition.

Tagging Body Repair Spots

Take a marker pen and walk around the body. Use the marker pen to circle the dents, holes, and other repairs that need to be fixed. There may be some holes you do not want filled in, such as bolt holes from the roll bar, seat brackets, or radio antenna (if you are replacing the antenna), etc.

When is it more cost effective to replace damaged body parts? Let's agree that anything can be fixed. For example, if you have a bent front fender, try to figure out how much time and money it will take to fix it and get a price quote from your body shop. Then, price out an original

A variety of alterations over the years will need to be reverted back to stock on this 1953 CJ-3A.

used fender in good condition. Original used parts are often better than the replacement parts available today. They fit properly and have the same metal thickness. Sometimes it's cheaper to replace a part than repair it.

Repair or Replace

Thoroughly inspect the body and then compare the price to repair the original body to the cost of buying a new replacement body. Reproduction Jeep bodies are available. If patching a rusty body costs nearly the same price of a new one, you have to make a choice. Purchasing a new body makes it faster and easier for you to finish the project.

If you go the route of a replacement body, it's critical that you keep the original body as a reference. Most of the holes need to be drilled or cut into the new body. Use the old body as a template.

If the damage is not too bad, it probably makes sense to repair the original body. Install high-quality patch panels if needed to get the best result. Trimming will be required to make the new panel fit your hole, but it will already be shaped correctly.

Body Filler

Many different and popular brands of body filler that people typically call "Bondo" are available. Bondo is the brand name of a common body filler. To keep it simple we will call all brands "body filler."

Over the years, I have seen many Jeep restorations that have used excessive amounts of body filler. If you are not familiar with this product, body filler can be used to cover up dents, scratches, drilled holes, and yes, even to fill bullet holes.

The process begins by sanding or grinding off the paint around the

To bring this Jeep back to its original look and condition, all of the previous modifications were corrected. Plates were welded in to fill both taillight holes and strengthen the body.

After the plates were welded in, they were ground down, and body filler was used to smooth the surfaces. When done correctly, the repair will not be noticeable.

This was the body that we decided not to use because of its advanced deterioration and rusty floors. When you do a restoration, never throw anything away. We used this body as a template for drilling the holes in the new replacement body.

Jeep replacement bodies are available. The quality and gauge of steel are good, and they were delivered in a wooden crate with no damage. It is often a good idea to buy the whole kit, which includes the tub, grille, hood, fenders, windshield frame, dashboard, and tailgate. This ensures that all of the body parts will bolt together properly.

damaged area. Then, by adding the right amount of hardener, the body filler changes to look like putty, caulk, or soft clay. Then, you apply it with a flexible plastic spreader and smooth it out as well as possible. When it dries or "sets," you can sand it smooth.

Most people doing restorations try to use the smallest amount of body filler as possible. They weld up the holes, work the dents out with hammers, and then grind the metal smooth. Only then will they put a thin coat of body filler to smooth out the imperfections. One of the reasons for not applying it too thick is that after bouncing and flexing, the body filler can crack, eventually blemishing the paint.

On the new body, drill holes for the seat brackets, gas tank hold-down strap, dimmer switch, gas tank fuel lines, and body-mount bolts. Most of this can be done easily by drilling the holes through the floor. The body-mount holes are tapped easiest from the bottom once you line up the body with the chassis mounts. Drill upward through the chassis bracket hole and through the floor.

Repairing a Dent

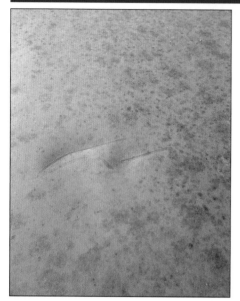

1 *Whether you're restoring an older Jeep body or using new body parts, there is always damage, such as dents, scratches, and small holes. Many of these can be repaired without using expensive tools or auto body shops.*

2 *Using a sander with medium-grade paper (80 grit) is good for this process to remove all the paint down to bare metal in and around the damage. Sand far enough away from the damage so that when you apply the body filler, you get a smooth finish. This also eliminates needing additional coats of body filler.*

3 *Try your best to get the old paint out of the deep scratches. If sanding does not remove the paint and rust from inside the scratches and dents, use an angle grinder and work the sides of the dents and scratches until you remove all of the paint.*

4 *Make sure to wipe the area clean with lacquer thinner before applying the body filler. This is very important in all phases of body work and painting. Before you apply any coats of body filler or paint the surface, it has to be clean. If not, the body filler, primer, or paint may not adhere.*

5 Mix the body filler and hardener with a 50:1 ratio on a clean, flat surface. A smooth piece of metal, wood, or cardboard will do. Make sure you continue mixing until you achieve an even color and texture.

6 Apply the first coat of filler with either a metal or plastic spreader. Read the instructions for curing time, as it can vary depending on the weather. Generally, body filler dries in 10 or 15 minutes, making it easy to sand.

7 When the first coat of body filler sets, sand it smooth with 80-grit sandpaper. Use a vibrating sander or block of wood wrapped in sandpaper to achieve a flat surface.

8 *Apply a second or even third coat, sanding between each coat to achieve a smooth, finished surface.*

9 *Use a fine 220-grit sandpaper for the final sanding.*

10 *Be sure to wipe down the finished area with lacquer thinner to clean off any sanding dust before priming. Use an automotive primer spray to protect the repair from rust.*

Small Holes

On smaller repairs, such as holes that held soft-top snaps and other holes, they can be brazed closed with a brazing rod. If you know how to use it, a small welder will also work on those holes. After the holes are welded shut, they need to be ground smooth, and if needed, a small amount of body filler is used to finish it off.

Tub and Floors

Jeeps for the most part are outdoor vehicles exposed to the elements. Jeeps typically don't rust much on the hood, grille, or front fenders. The most common ailment is rusted-out floors in the tub (the main body). Both the driver and passenger floorboards get lots of use. The paint erodes in a short time due to people's feet tracking in mud and dirt. Then, it gets ground around. The bare metal begins to rust when exposed to the elements. All four tires are constantly throwing mud and snow up against the bottom of the Jeep as well.

It's common practice when preparing a Jeep body for paint to not repair one scratch or dent at a time. Grind and sand several dents and scratches at the same time. Then, mix enough body filler to smooth that area. Sanding and priming are next.

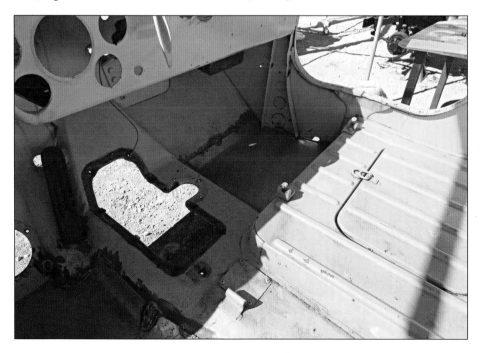

The nature of the Jeep, being an outdoor vehicle for both work and recreation, exposes it to the elements. The most common place for deterioration is the floors. Your two options are to repair the floor panels or to replace them. If the floors are not rusted through, replacement floor panels are available to patch areas with pitted metal. Replacing the full floorboards takes some custom fitting to make them fit right. After they are welded in place, the welds are ground down and body filler can be used to hide the seams.

Repairing and Replacing the Floors and Braces

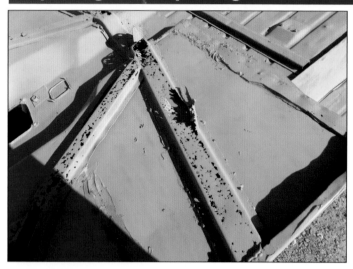

1 The channels on the bottom of the Jeep are known as braces. These braces not only give the bottom of the Jeep strength but they also tend to hold moisture, and they are the first thing to rust out. Replacement braces are available, and if the floor is not in bad condition, the braces are easy to replace.

2 Using an angle grinder, simply cut the spot welds that hold the brace to the bottom of the floor. Some Jeeps have braces that are so deteriorated that they can be removed with a hammer and chisel. Avoid using a cutting torch because the heat can warp the floor.

3 After removing the old braces, clean up and prepare the bottom of the Jeep to accept the new braces. Grind down any old welds. Set the new braces in place and tack weld them every few inches.

4 *Once you have replaced all of the braces and repaired the floors, get a good coat of primer on this new metal. Depending on your climate, surface rust can form immediately.*

5 *If done correctly, repairing the floors and replacing the braces make the bottom of your Jeep look original. Once the bottom of the Jeep body is painted and the body is replaced on the chassis, people will be impressed when they look under your Jeep.*

6 In a more dramatic restoration where the floors are beyond repair and have to be replaced, the first thing you have to do is remove the toolbox. This can be done by drilling out the spot welds or using a cold chisel or angle grinder, whichever one causes the least amount of damage to the toolbox.

7 You can try this on your own, but an experienced welder who is also a steel fabricator will usually do a better job. Here, we cut out the defective floors leaving metal around the edge so that the new floors are well supported.

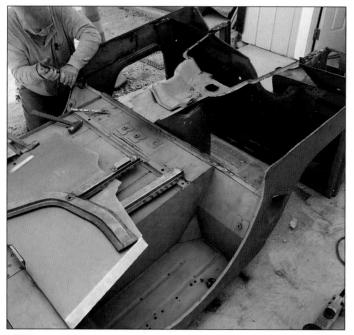

8 Here, a tool called a Cleco was used when reassembling the floors and the braces. Start by drilling a series of small holes and inserting the Cleco to hold the two pieces of metal together. Then, spot weld everything together, remove the Clecos, and weld in the remaining holes.

9 When the new floors were installed, it was a work of art. It takes someone with skill and talent to do this properly. When you factor in the cost of the material and skilled labor, many times it would be less expensive to buy a replacement body.

There are times when it is cheaper and faster to replace the old body with a new reproduction body. The result will be better than piecing together a bunch of panels and patches. You won't need a tetanus shot when preparing a new body. However, new bodies are somewhat universal and generic. They are designed to be used with a variety of transmissions, lights, etc. So, there will be some drilling and cutting to make the new body fit your specific Jeep.

Preparing the New Body

1 *When you order these replacement bodies, you get a little and give a little. You get a nice, straight rust-free, clean body, which will add to the quality of your restoration. However, you have to drill many of the holes needed for lights, seats, and other accessories.*

2 *The firewall in this new body was set up for different master cylinder applications. This 1965 CJ5 had a heater system when it was manufactured. We had to modify this body to accept the heater and many of the access holes, such as for the parking brake cable.*

3 *Use the old body as a template. Remove the paint and dirt from the firewall, and then cut out the metal piece that fits the accessories and cables you need. Use an angle grinder to cut the straight lines.*

4 *Clamp the old firewall section over the new firewall and trace it. Use a fine-tip marker or a razor knife, which will leave a finer line.*

5 *Use an angle grinder to follow the lines, keeping them as straight as possible. The old firewall piece and the new firewall are the same gauge of metal, so it should fit properly, making it easy to weld.*

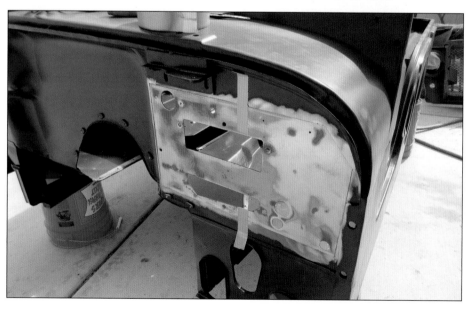

6 *When replacing metal parts like this, if you can use clamps to hold them in place, do so. In this situation, that wasn't easy, so duct tape was used. Once the corners are tack welded, the new piece is held in place and the tape can be removed.*

7 *For this gauge of metal, an inexpensive 120-volt wire-fed welder will work. When the welding is complete, it will be strong and look presentable.*

8 *Spot welding leaves about 1 to 1½ inches of gap between each weld. This prevents warping. Continue to weld between the spot welds until the seam is completely welded.*

9 *After the welding has cooled, take an angle grinder and grind off the extra weld bead, making it as smooth as possible. In some applications, it is necessary to do on both sides for aesthetic reasons. Be careful not to grind so deeply as to leave marks on the metal or grind out the weld you just made.*

10 *After grinding is complete, the weld looks very presentable, and in some cases, you can just prime and paint over this finished product. In visible areas, it may be necessary to use body filler and finish off the seams so that it looks like an original body part. This is a judgment call depending on the location of this type of repair.*

11 *Marking the location of the new holes you have to drill is easier if you use the old body as a template. This can also be done with photographs and looking at other finished Jeeps. Most holes can be drilled with a traditional drill bit or a step bit (like the one being used here) to widen a hole.*

Toolbox

The toolbox under the passenger's seat and the underside of the gas tank are traps for moisture in older flat-fender Jeeps. Under the floors, there are narrow channels, or ribs, that give the floors support. They are called braces. Body mounting bolts go through them, creating places for the moisture and mud to collect.

Replacement panels for both floorboard panels and braces are available for most models. Keep in mind that to replace the floor panels and the braces, you have to remove

It is very common in most Jeeps to have rust inside of the toolbox. If the passenger-side floor is okay but the toolbox is rusty, there are several ways to handle this. Cut two or three pieces of sheet metal, insert them through the toolbox cover, and weld them together from the inside. However, the inside of a toolbox is dark, and there is not much room to maneuver the welding gun.

The best way to repair the inside of the toolbox is to remove the toolbox itself and replace the entire passenger-side floor with new metal. A one-piece panel will cover the passenger-side floorboard and the inside of the toolbox, giving it a better look.

From 1946 to 1986, there were a few styles of tailgates. The 1946–1971 Jeeps have "Willys" stamped in the center. Military Jeeps and 1972–1975 CJ models were blank. From 1976 through 1986, the tailgate had a larger "JEEP" stamp. Here, there are holes drilled into two of these tailgates for jerry can brackets. You have many choices in handling those holes. Some people like to place a nut and bolt with a decorative washer into the holes and then paint. This leaves the hole available for future owners to add a jerry can. If you would like to fill the hole, follow the same procedure as with auto body repair and welding a hole closed, grinding it smooth, adding body filler, and finishing. It's common to have rust on the bottom edge of the tailgate. If there is substantial rust, don't bother refinishing it, buy a new tailgate instead.

the body from the chassis. When cutting out the old braces and floors, try to use an angle grinder to cut the old welds. A cutting torch creates too much damage to the adjoining metal and weakens it. I can tell you from experience that, generally, reproduction floor panels fit perfectly. After you're done with the restoration and someone looks underneath, you'll be happy you replaced them.

Tailgates

The military Jeeps (MB/GPW, M38, and M38-A1) did not have tailgates, although some people cut the rear area out and added tailgates. All CJ models do have tailgates.

The 1946–1971 Jeep CJ models have the hanging chain set that makes it easy to drive the Jeep with the tailgate up/closed or also in the down position to haul cargo. If you are driving with the tailgate in the down position, be aware that the tailgate hinge is slotted for easy removal. Likewise, if you are driving and hit a bump and the tailgate bounces up, it can come off the hinges.

The 1972–1986 Jeep CJ models had cables supporting the tailgate when it was open.

It is common to see older Jeeps with lots of rust on the tailgate. The lower part of the tailgate has a curve to it to hold the hinge rod, and it catches lots of rainwater. So, you see many of them with a row of rust on the lower tailgate.

You can purchase new replacement tailgates. Be aware that there are two different versions. An unlicensed version does not have "WILLYS" stamped into the tailgate. It is blank. These are less expensive.

The officially licensed tailgate does have the "WILLYS" name stamped into it. If you just need a generic tailgate, get the cheaper one, but if you're doing a complete restoration, you need the one with the name "WILLYS" stamped on the tailgate. That is the only difference. They are the same otherwise.

Metal Patches

Many Jeeps have long histories and lots of stories behind them. They may have had holes cut or drilled into them that need to be filled. Radio and speaker holes in dashboards are examples.

If you want to bring the Jeep back to its original state, do away with them. You can trace the shape of the existing hole on a piece of thin cardboard and cut a metal plate of the same shape. Next, weld it in and then grind down the welds to make it smooth. Finally, use a thin coat of body filler to hide the seams. If done correctly, you can't tell it was ever there.

Repairing Metal Damage

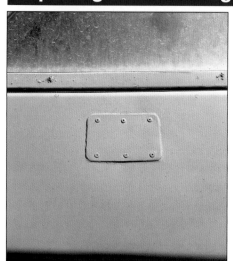

1 *After removing this riveted patch, we uncovered and identified a hole in the windshield frame. This will be repaired with the same procedure as for a radio hole in the dashboard or an oversized taillight hole.*

2 *Use an angle grinder to grind the heads off the rivets, making it easy to remove the patch. Also, grind the paint and rust from around the hole to make a clean area for the repair work.*

3 *Cut a metal plate larger than the hole being filled using a piece of either 18- or 20-gauge steel. Then, place the plate over the hole and use a pencil or chalk to trace an outline on the body over the hole.*

4 *Use a 4½-inch angle grinder with a cutting disk and follow the outline, cutting out the square hole in which the new plate will be installed. Be careful to follow the lines and not to overcut the corners. You can use a medium file to remove any burrs where you have just cut.*

5 *Remove any paint, old body filler, and dirt from the area surrounding the hole to be repaired. Use a vibrating sander or sandpaper wrapped around a block of wood. A medium-grade 80-grit sandpaper will do.*

6 Use a 110-volt wire-fed welder to weld in the patch metal. Be sure to use a heavy-duty extension cord. You can hold the metal in place with a large C-clamp while you weld.

7 Do not weld a continuous bead around the patch. That would create too much heat and warp the metal. A series of small spot welds are enough.

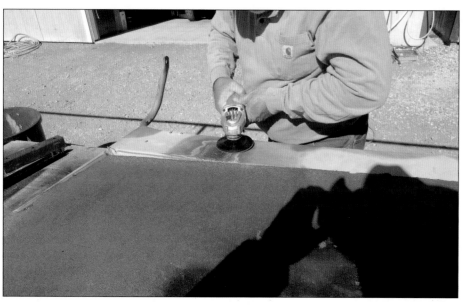

8 In a few moments, the welds will cool, and you can grind them smooth with the angle grinder. Keep in mind that with most dashboards and fenders, you don't have to finish the back or the underside of the fender because no one will see it. But on the windshield frame, both sides have to be finished.

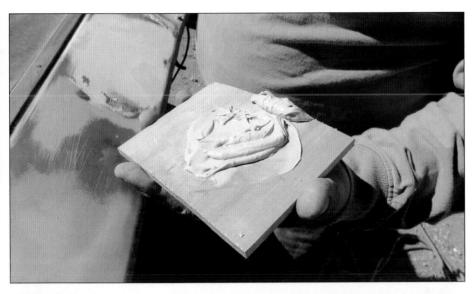

9 Once the welds are ground smooth, it's time to apply body filler. Make sure you have sanded all the paint around the area of the patch. Wipe the area clean with a rag soaked in lacquer thinner to remove any dust before applying body filler.

10 Mix the body filler with a 50:1 ratio of putty and hardener. Hardener comes in many colors. Mix it thoroughly so that the color is even. Use a metal or plastic spreader to apply it evenly. After the first coat has dried, sand it smooth with 80-grit sandpaper. Repeat this process two or three times and finish with a fine sandpaper, such as 220-grit.

11 Use an automotive primer from an automotive paint supply store. This protects the bare metal and aids in the paint adhesion to the Jeep body. For larger areas or priming the entire Jeep, use a spray gun.

Body-On Restoration

If you're not doing a frame-off restoration, you can take the Jeep apart mechanically. Remove the engine, gas tank, wiring harness, and anything else that might be in the way at the paint shop. I have the shop do all of the body work and just prime it. Then, I paint the frame in "chassis black" paint and put it back together with a freshly rebuilt engine.

Once the Jeep is running, I road test it to make sure everything is working (brakes, electrical, wipers, etc.). Then, I bring it back to the body shop for a paint job. Once I get it out of the paint shop, there is nothing to do other than install the seats and top.

Paint

Keep in mind that most vintage Jeeps were painted with single-stage paint (often an enamel paint).

Acrylic urethane paint is more common now for basic solid-color paint jobs. It is very easy to apply and is very durable. In the early 1980s, some CJ models had metallic paint and used the modern base coat/clear coat method.

Once you or the auto collision shop paint your Jeep, it's time to put it back together. This process should be the reverse of the disassembly we covered in Chapter 4.

Military Paint Colors

Army Jeeps from the World War II and Korean wars were painted with different shades of Olive Drab Green. You will also find that different branches of the military used different colors. The U.S. Navy often used Gray, etc. The paint finish often had a flat or a satin sheen to reduce reflections that could be seen by aircraft. Companies that specialize in selling paint for military vehicles have what you need and can tell you the correct color.

The engine, body, and chassis are all the same color, and you can paint your wheels in the body color as well. Most restorers leave the rubber around the windshield glass and the rubber windshield-to-tub gasket unpainted. Under the hood, the coil, cap and wires, plugs, fan belt, radiator hoses, fuel lines, carburetor, fuel pump, and a few other things are left natural (not painted). With the "flat white" stencil markings and "blue drab" hood registration numbers, the only contrast is the color of the seat cushions.

Painting a Military Jeep

1 *Once the body work and priming is done, a final sanding is needed. A simple vibrating sander and 220-grit sandpaper will get the job done. Wipe off all dust and prime the entire surface. Another light sanding with the 220-grit sandpaper may be needed after priming.*

2 *Look for any parts you might have overlooked that need sanding. Inspect each part for any dents or body work you may have missed. Also, inspect for dust and dirt in the primer. Then, hang as many body parts as you can from a rack so that you can paint them on all sides.*

3 *Make sure to clean all of your glass parts, including windshields, gauges, and headlights, before you mask them off. The tape will stick better to a clean surface. Also, when your paint job is completed and you remove the masking, you don't have to clean the glass with harsh chemicals, which might discolor your new paint.*

4 *Oddly shaped body parts, such a seat frames and top bows, have to be hand sanded using a 220-grit sandpaper.*

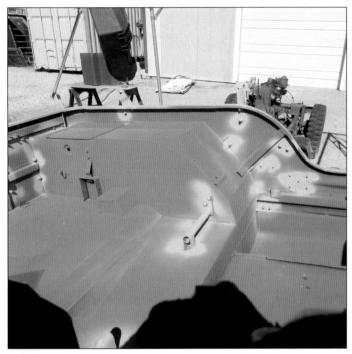

5 Be sure to prime all the new hardware before painting it. You can spot prime these items with automotive primer from a spray can, or if you prefer, use your spray gun to re-prime the entire Jeep once the hardware is installed.

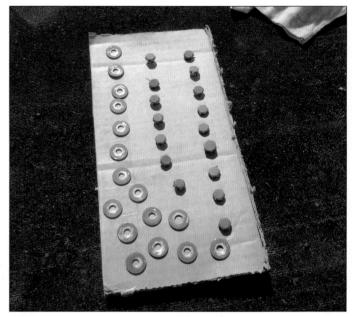

6 A few things have to be assembled after the paint job is complete. In this case, the body will be installed after being painted, and the floorpan will be bolted in. Use a piece of cardboard and stick the bolts through a hole to keep them in place for easier painting. As always, inspect the bolt's threads to be sure they're reusable, if you haven't replaced them already.

7 This single-stage military paint had a formula of its own. Each manufacturer has a recommended thinner for its paint. The label on my can recommended using Xylene as the thinner, which is available at most large hardware stores.

8 Make sure to pour the paint through a strainer to remove any lumps or debris. With this paint, you only need about a 5-percent ratio of the Xylene thinner to the paint. Mix only enough for one canister at a time for the spray gun.

9 *Always start by painting the bottom of the Jeep. Anyone painting needs to wear a respirator, and anyone helping them needs a mask. Safety first. With this paint, you need a regulator on your spray gun or your compressor set to 29 psi.*

10 *Don't be worried if you miss a few spots. Take some of the paint to your local auto paint supply store so they can put some in spray cans. It's always great to have extra paint in spray cans for future touch-ups or replacement parts.*

11 *Once all of your parts have been painted and dried, you can assemble the Jeep. Assemble in the reverse order that the Jeep was disassembled.*

Civilian Paint Colors

Most owners want to stay with original Jeep paint colors. You can look up what colors were available for the model year you have. The best place to see the original color is to take a flashlight and look inside the toolbox under the passenger seat. Some of the best Jeeps were not original colors. It depends on what you're going for. If it is an original restoration, stay with the factory colors. If it's a creative project or an off-road toy, pick any color.

Painting the Wheels

Be aware that the lug nuts on the driver's side of 1941–1971 military and civilian Jeeps are left-hand thread. This is the opposite of what most people are used to. If the lug nuts don't seem to want to come off, try turning them the opposite way.

Next, make sure the wheels are straight and run true. Jack each wheel off the ground and spin it looking for wobble. An old trick is to hold a screwdriver against the body and spin the tire to see if the gap changes.

If the wheels are straight, have a local tire shop remove the tires from the wheels. Call ahead if you have old tires. Some shops will not remove tires more than 10 years old due to liability. Tires manufactured after 2000 have a date code molded into the sidewall. The last two digits are the year.

Depending on the condition of the paint on the rims, they may need to be sandblasted. Sandblasted or not, make sure they are clean and prime them. You can paint them with a spray can or have the body shop paint them. Then, take them to a local tire shop and have the new tires mounted and balanced. If you want a long-lasting finish, you can also powder coat them, but that can be expensive.

Having an extra set of old wheels and tires to use during the restoration saves the freshly painted wheels and new tires from grease or overspray. Just swap them when you're ready.

Single- Versus Two-Stage Paint

A single-stage paint job has the paint and clear coat all in one layer. A two-stage paint job is just that: two layers of paint. The first layer is the base paint color, and the second layer is the clear coat that makes it shiny. The cost varies depending on the shop that is used. Sometimes one is cheaper than the other and sometimes vice versa.

Painting

I learned a long time ago that you can't beat a professional at his own trade. Just like the engine machine shop, most of us can't attempt some of the highly trained trades and put out a quality product. Now, if you own a body shop or are an experienced auto body painter, have at it. You will probably give a lot more care to prepping your own Jeep and do a great job. For the rest of us, I do not recommend attempting the paint job. Having a paint booth takes space and costs a fair amount of money.

When refinishing wheels, as with bodies, some are beyond help. If the wheel is very rusty, you may need to have it sandblasted or replace it. Some 80-grit sandpaper and a few minutes of your time will get the dirt and surface rust off the wheel and ready for primer and paint. It is a lot better to dismount the tire from the wheel before prepping and painting, especially if you plan to replace the old tires. However, you can mask off the tires with tape and masking paper, then use a spray can to paint your wheels while the tire is still mounted. However, you can't see or paint the inner part of the wheel with this method.

Also, it takes skill to properly mix the paint, set up the paint gun, and spray the paint in an even pattern.

Reassembly

Make sure you order a new set of body mounts. The kit comes with all of the rubber insulators and mounting pads you need to reinstall the body to the chassis.

When the chassis is ready and the axles and drivetrain have been installed, carefully mount the body back onto the frame. Having four or five strong helpers allows the body to be aligned with the body mounts and lowered gently. It's helpful to have a person or two underneath to help in aligning the body tub to the body mounts. Once lowered, the body can be wiggled around as necessary to install the body bolts.

Bolting the tailgate, windshield frame, front fenders, hood, and grille is straightforward if you are reusing the original parts. A bit of aligning will probably need to be done. Don't tighten any of the bolts until the body is completely assembled. Once everything lines up, you can tighten the bolts to spec.

Fenders

Most fenders are held in place by three bolts from the fender to the tub and from the fender to the grille. Some models have a bracket running down the underside of the fender bolting it to the chassis.

Grille

All vintage Jeep grilles are bolted to the chassis by a single bolt in the center of it and bolted to the fenders on their leading edge. The large bolt in the center of the grille is very important because it holds the grille

Not much has changed since 1941. Both flat-fender and round-fender Jeeps have been put together over the decades the same way. Depending on the model, three or four bolts hold the fenders to the main body. Most have threaded nuts inserted into the body.

The same method for mounting the fenders to the body is used for mounting the fenders to the grille. The leading edge of the fender attaches to the grille with three bolts.

This inside fender bracket is not found on all models, but if you have one, it should be attached to the frame. The bolt's exposure to the mud and salt can cause it to be so rusty that when you try to remove the fender, the bolts break. Replace the bolt and the bracket if they've broken during disassembly.

A large bolt holds the bottom of the grille to the chassis. Look underneath the chassis to make sure there is a body mount grommet and washer. This prevents the front clip of the Jeep from squeaking. This bolt holds the front clip to the chassis.

In addition to the center bolt, the Ford GPW and the Willys MB, CJ-2A, CJ-3A, and CJ-3B models had mounting pads on the frame under each end of the grille.

TECH TIP

Installing as an Assembly

Many restorers and mechanics find it easier to bolt the radiator to the grille and then install the grille to the Jeep with the radiator all in one piece. This makes things easier if the fan blade has been installed on the water pump and the Jeep has a radiator shroud. ■

To save time and not damage the radiator and grille, install the grille and radiator as one unit. Then, you can reconnect the upper and lower hoses, install the grille support rod(s), and install the fender bolts on each side of the grille. Reconnect the headlight and turn-signal wiring. Install the big body-mount bolt on the bottom of the grille. Finally, refill the radiator with a 50-50 mixture of coolant and replace the radiator cap.

and the fenders to the chassis. If this bolt is not installed securely, the fenders and grille will rattle.

In addition to the center bolt, the Ford GPW and the Willys MB, CJ-2A, CJ-3A, and CJ-3B models had mounting pads on the frame under each end of the grille.

Welt

Welting is a canvas band added to the top of the grille to give the hood a resting spot and to deaden the metal-on-metal sound. You'll want to replace the welting and rivets on your Jeep's grille.

Grille Support Rods

On early military and CJ flat-fender Jeeps, only one support rod runs from the firewall to the top of the radiator. In the later round-fender CJ models, two grille support rods run from the firewall to the top of the grille shell.

Call it weatherstripping or welting, but whatever you call it, it is necessary to install this on top of the grille on all models from 1941–1971. It serves to keep rain, snow, and dust from entering the engine compartment, and it prevents the metal from the hood and the grille from making contact. Line up the welting with the holes in the top of the grille and use a simple leather hole punch to add holes as necessary. The hole punch will easily go through the woven fabric.

Using a simple pop-rivet gun that is available at any hardware store, find the appropriate-size rivet that fits the hole in the grille. Buy an assortment of sizes and find the one that fits. Don't be afraid if you install the wrong-size rivet, they are made out of aluminum, and you can easily remove them with a hammer and chisel.

After you have installed the welting, it offers a nice, finished look to the top of the grille. Some people like to leave it natural, and others paint it to match the body.

Welt is needed on the windshield brackets where the hood rests when it's open. Cover these to protect the hood from scratches.

Windshield

There are three reasons for removing the windshield glass from your Jeep. One is that it is broken or discolored (showing air bubbles and discoloration around the edge from the glass layers separating) or it has stone chips and cracks. The second reason may be that your glass is an excellent condition, but the rubber gasket around the edge is dry rotted or cracked and you would like to replace the gasket. The third reason is to paint the windshield frame. In this instance, removing the glass makes for a better paint job. In most cases, you have to replace the rubber gasket too.

Two grille support rods run from the top of the grille to the firewall on M38-A1s and CJ-5, CJ-6, CJ-7, and CJ-8 models. These bars prevent the grille and radiator from rattling, thereby preventing damage and adding strength to the front clip of the body.

In the earlier Willys MB, CJ-2A, CJ-3A, CJ-3B, and M38 models from 1941 to 1964, only a single support rod goes from the firewall to the top of the radiator.

Removing the Windshield

The Willys MB/Ford GPW and CJ-2A built from 1941–1949 used the opening style of windshield that is shown in the Jeep in the background. The CJ-3A through the later CJ models (1949–1986) used fixed windshields as seen on the Jeep in the foreground.

1 To start the process of glass removal, remove the windshield frame from the Jeep. Work on a nice, flat table covered with a clean blanket or other padding so as not to scratch the glass.

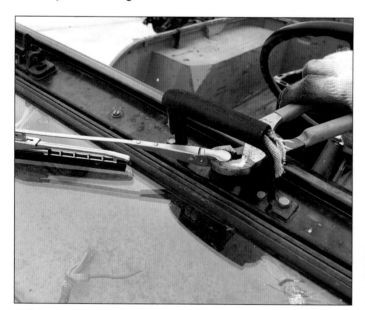

2 Remove the windshield wiper blades and wiring/vacuum lines from inside the windshield frame. If there are any other items attached to the frame, remove them now as well.

3 Remove the small rubber strip in the center of the gasket, which is inserted to tighten the gasket. Simply use a screwdriver to pop out the end and then pull it out.

4 *Depending on the amount of exposure to the elements, the strip may come out in one piece or in chunks.*

5 *After the tightening strip has been removed, use a razor knife to cut the top half of the rubber gasket. This exposes the edge of the glass and makes it ready for removal.*

6 *Exposing the glass edge allows you to lift it out easily.*

7 *Remove the glass from the early pop-out windshield frames by loosening the nut and bolt in the upper right- and left-hand corners of the windshield. Once this bolt is removed, the channel holding the glass comes apart. It may need a little finessing due to years of rust and paint. Remove the upper channel of the frame, slide the glass out, replace it, and put the channel back. A rubber gasket is available that sits in the channel and holds the glass. If you cannot locate a gasket, clear silicone caulking can be used. At the same time, add new weatherstripping on the outside around the windshield.*

9 From 1949-on, the rubber-gasket windshield system was used. If your glass is cracked, broken, or faded and you need to replace it, follow this procedure. If your glass is good and you want to reuse it and your gasket is dry rotted and cracked, remove the glass safely in one piece so you can repaint the frame and replace the rubber gasket. Use a screwdriver or cotter-pin puller to remove the rubber key in the slot.

10 After the key has been removed, use a utility knife to cut down the gasket halfway exposing the glass. Have someone assist you so that when you cut most of the gasket away, someone can hold the glass and it doesn't fall out. Keep the glass in a safe place while you are finishing the rest of the restoration. New glass can be purchased online or cut by your local auto glass shop.

11 By turning this wing nut, you can open or close the windshield for ventilation in various positions. You can also remove the nut and take the entire windshield glass frame out of the windshield itself.

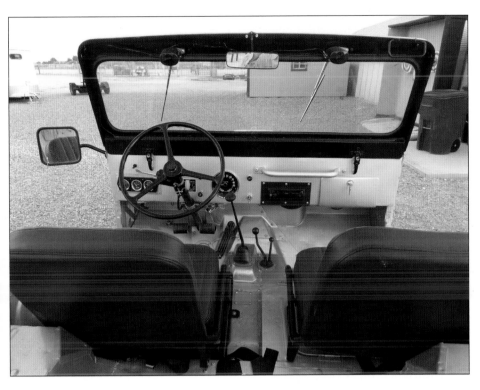

Early Jeep models (1941–1975 MB, GPW, M38, M38-A1, CJ-2A, CJ-3A, CJ-3B, CJ-5, and CJ-6) have two windshield clamps on the top of the dash. When released, the windshield lays down on the wooden hood blocks. A web strap is used to secure the windshield to a footman's loop on the grille.

Have someone help you remove the windshield so that the glass will come out in one whole piece if your intentions are to save and reuse it. Store the windshield in a safe place because the glass is expensive. If it is already damaged, it doesn't matter if it comes out in one piece. Early Jeep windshield frames have two latches on the dash that hold the windshield upright.

On the 1976-and-newer CJ-5, CJ-6, CJ-7, and CJ-8, two knobs thread through triangular plates on the dash to hold the windshield vertical.

Replacing Windshield Glass

Vintage Jeep windshields all have flat glass. Before you replace the windshield glass, make sure you have a new windshield gasket too. If your windshield has been out in the sun for a few years, the rubber gasket can be dry rotted, and it will come apart when a shop tries to remove it. New windshields and rubber gaskets can be ordered online or possibly from your local auto glass shop.

When you take your old windshield to a local auto glass shop, they might be able to cut new glass and replace the windshield themselves. However, not all glass shops have the talent or know the tricks to cut flat glass. They usually replace modern curved windshields, so call them ahead of time before bringing your flat windshield to them.

Later CJ models (1976–1986 CJ-5, CJ-7, and CJ-8) use a large knob on each end of the windshield frame to keep the windshield upright. Removing the two knobs allows the windshield to lay down on the footman's loops.

Restoring Your Windshield

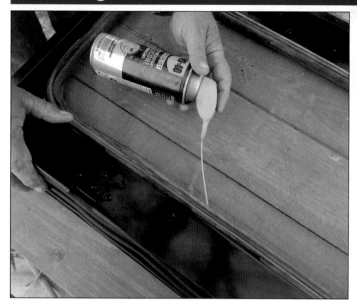

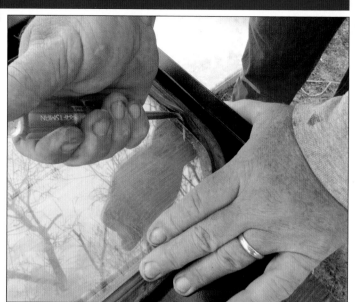

1 After you have set the gasket in the slot on your windshield frame, spray WD-40 or WD-40 silicone in the windshield groove for lubrication. This helps you install the glass.

2 Slide the glass in the rubber track starting at the bottom of the windshield. Gradually, as your helper holds the glass, use a cotter-pin puller (or small screwdriver) to pull the rubber lip out to cover the glass.

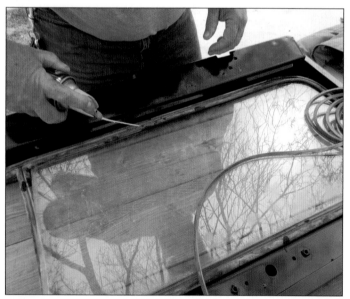

3 Before you completely install the rubber all the way around, install the key in the slot in the bottom of the windshield. This keeps the glass from sliding down into the rubber gasket and collapsing the key slot. Many people set the entire glass in the gasket and then set the entire key. It can be done either way. Installing the key on the lower side first gives structure to the gasket and doesn't allow the glass to slide around. To make this job easier, spray WD-40 in the key groove. This lubricates the rubber.

4 Inserting the key in the gasket is not difficult, it just takes some strength and persistence. Pry open the slot in the gasket with a pick or small screwdriver, follow up with your thumb pushing the key into the slot. You can also do this with the rounded end of a screwdriver handle. Be careful not to push too hard on the glass side of the slot; you don't want to put too much pressure on the glass.

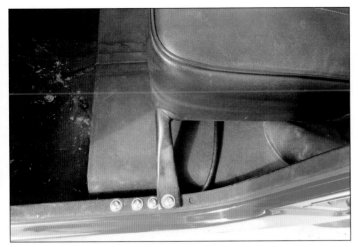

Military and civilian Jeeps built between 1941 and 1971 have the gas tank under the driver's seat. Some have one metal strap and others have two to hold the gas tank in place. When installing the hold-down straps, use new felt or rubber insulators to prevent the hold-down straps from rubbing on the metal gas tank.

Dash Installation

If you are lucky enough to have a 1976–1986 Jeep with a bolt-on dash, you can reinstall the gauges and stereo speakers now. Then, simply bolt the dash to the body tub and reattach the wires you labeled when you took the Jeep apart. Adding a new glove box may be a good idea too. There will be a few cables for the vent and heater, the speedometer cable, and the passenger grab bar, but overall, it's not too bad.

Next, carefully slide the steering column through the hole in the dash and firewall. Then, reconnect it to the steering shaft under the hood. Under the dash, snap together the column's electrical wire plugs and install the two bolts that hold the column up. Then, add the plastic column cover around the column on the face of the dash. Hopefully, you checked the steering column's turn-signal switch, ignition switch, and horn before you removed it. If so, they should be working now.

If your dash doesn't unbolt, you need to throw down a moving pad and lay on your back to reach under the dash. The process is similar but less comfortable. Reinstalling the gauges and reconnecting the wires is pretty straightforward. The steering column attaches to a clamp under the dash with two bolts. A wire inside the column hooks up to the horn button. Reattach the parking brake handle and cable. Then, reattach the ignition switch. Riveting the brass dash plates back on is pretty easy too.

Steering Wheel

The next step is to reinstall the steering wheel if you removed it. You may want to delay this step until the gas tank and driver's seat are in place in older models. Waiting could make the seat installation easier. Once you are ready to install the steering wheel, line it up as well as you can and try to push down on the spokes. Be sure the horn wire is coming up through the center. You will probably want to lubricate the steering column splines a little and you may need to tap on the steering wheel hub a bit with a rubber mallet. Aftermarket steering wheels may require an adapter kit. Once you tighten down the steering wheel, attach the horn's wire and install the horn button. Then, test the ignition switch, turn signal, and horn to be sure they still work.

Fuel Tank

The 1941–1971 military and civilian Jeeps have the gas tank under the driver's seat. Reinstall the tank before installing the seats. When installing the hold-down straps, you'll want new felt or rubber insulators to prevent the hold-down straps from rubbing on the metal gas tank. Next, reconnect and tighten the hold-down straps.

You'll also need to reinstall the tank's drain plug from underneath the body. Finally, reconnect the fuel line. If you have a civilian Jeep, reconnect the electric wire to the sending unit on top of the tank.

If you have a 1972–1986 Jeep CJ, you can hang the gas tank under the rear cargo area any time after the body has been painted. Two straps suspend the tank, and it will have a skid plate for protection. Connect the fuel line and sending-unit wires before lifting the tank into place. Be sure to connect the rubber filler and vent hoses. Replace them if they show any signs of cracking or hardness. When replacing hoses, use new hose clamps.

Seat Frame and Cushions

From 1941 through 1971, Jeep seat frames and cushions didn't change much. Although, it's always best to research the year and model of your Jeep to determine what the seats should look like. Unbolting them is as simple as finding their bolt locations in the floor and underneath.

Driver's Seat

Early Jeeps had various configurations on the driver-seat frames and mounting brackets. They were bolted solid to the body for safety reasons and to protect the gas tank. On the 1941–1945 Willys MB and Ford GPW, the lower driver's seat cushion flips up to reveal the filler neck on top of the gas tank.

On 1976-and-later CJ models, the driver seat bolts onto seat-riser brackets that allow the seat to slide back and forth.

Passenger's Seat

Willys MB and Ford GPW seats were bolted directly to the floor with a pivot hinge and no toolbox underneath. Jeep CJ models from 1946 through 1971 had a toolbox under the seat and pivoting seat hinges with a quick release.

On later CJ models, the passenger's seat bolts onto seat riser brackets that allow the seat to slide back and forth. Some tilt forward.

Seat Cushions

There are many different varieties and colors from which to choose. If you're going for a classic authentic restoration on a military vehicle, purchase new Olive Drab canvas covers. They are easy to attach.

Vinyl Seat Cushions

A local auto upholstery shop can

The 1941–1945 MB and GPW passenger seats are bolted directly to the floor of the Jeep because there is no toolbox under the passenger's seat. When removing this seat, unbolt the bracket from the floor, leaving it attached to the seat frame. This exposes any rust or damage to the floor so that it can be repaired.

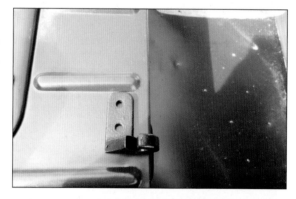

From 1946 to 1971, this was the conventional passenger-seat bracket with a slip hinge. When installing on a new body or relocating, place the bracket on the seat frame, put the seat in the position you want, and trace the holes in the top of the toolbox. Remove the seat bracket and drill out the holes and then replace the bracket. Finally, bolt the brackets to the top of the toolbox.

This seat frame and bracket are designed for easy removal. Just tilt the seat forward to about a 45-degree angle and lift straight up. To replace it, do the reverse.

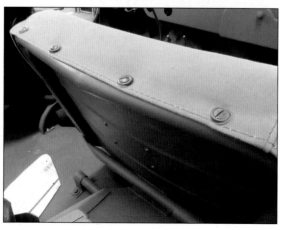

Attaching seat cushions to their respective frames is very easy. The Willys MB and Ford GPW cushions are easily attached with a sheet-metal screw and decorative washer. They screw into the seat frame where there are predrilled holes. This system worked so well that it continued through the CJ-2A, CJ-3A, and CJ-3B models.

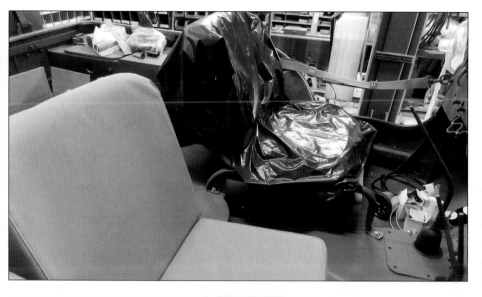

After the cushions have been attached and the seats are installed, keep them clean and covered until the restoration is complete. You can simply use a large 30-gallon plastic garbage bag to cover the seats.

Many materials can be used to reupholster the seats. Vinyl seems to be the most popular and durable to withstand exposure to weather. Vintage Jeep cushions are attached to a flap with screws into the tubular seat frame. Some upholsterers make a one-piece top cushion that slides over the back. This bottom cushion has a piece of plywood in the bottom and is bolted to the lower seat frame.

Some newer Jeep CJ models are upholstered with patterns and have bolsters for more comfort. A good auto upholsterer can re-cover this seat following the same pattern and designs. These seat covers will be attached to the foam permanently by the upholsterer. Sand and paint the seat frame legs that stick out before you send it out to be reupholstered.

rebuild or recover your civilian Jeep seats. Vintage CJ models from 1946 through 1979 have low back seats. The 1980-and-newer CJ models have high-back seats. Seat material from the 1960s, 1970s, and into the 1980s have patterns and designs in the material.

Rear Seats

Rear seats were not generally included when vintage civilian Jeeps were new, but they could be added as an option. Many older Jeeps do not have them, but reproductions are available. They attach the same way as the driver's seat (by being bolted to the floor). The one exception is regarding the World War II rear seats used in the Willys MB and Ford GPW. Their frame fits in a slot on either side of the rear wheel wells. They fold up for extra space.

Miscellaneous

Now is a good time to replace all of the weatherstripping. Be aware that the new rubber or foam pieces will be thicker than the old ones. This can affect how the parts fit. The weatherstripping will squeeze down over time with use.

Next, reinstall the bumpers, the frame cover ahead of the grille, the license-plate bracket, the gas-filler grommet or bezel, side mirrors, hood catches, headlights, taillights, side-marker lights, rubber fender flares, and decals.

CJ-7s and CJ-8s may have aluminum rocker trim that needs to be replaced or reinstalled. The roll bar, spare tire carrier, and spare tire will need to be reinstalled too. Also, the cowl vent cover needs to be installed when the paint is dry. Finally, the small interior pieces can be reinstalled.

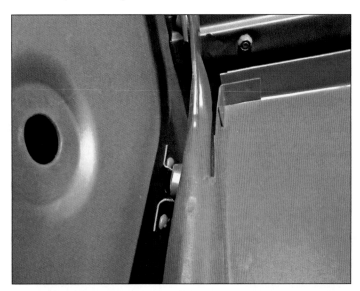

The 1941–1944 MB and GPW rear seat fits easily into slots on either side of the Jeep, allowing the seat to pivot and fold up for more cargo space when not transporting people. Two hold-down hooks on the rear seat back hold the seat to the body and allow it to slide up and down while folding. After this frame is painted to match the Jeep, the seat cushions attach the same way as the front seat cushions with the screws and washer on the outside edge. However, on these reproduction seat frames, you may have to drill the holes for the seat cushion screws.

WIRING HARNESS

The wiring harness in your Jeep is probably the most complex and misunderstood system. There are many stories of people who had a Jeep with electrical shorts in the wiring that couldn't get the lights or accessories to work. So, they took it to a shop and spent hundreds of dollars on diagnosing the problem when for the same price, or less, you can buy a new wiring harness and install it yourself.

If you think about it, the wires in your Jeep are anywhere between 35 and 80 years old. The insulation on each wire is brittle and may be crumbling off in places. In addition, previous owners may have been

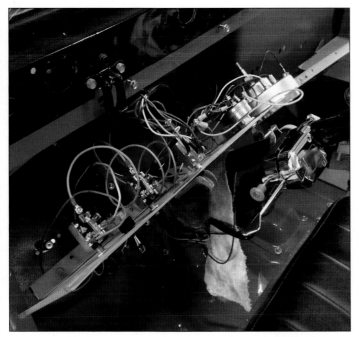

Round-fender Jeep CJ models from 1955–1986 have a bolt-on dashboard. This makes it easier to replace the wiring. Flat-fender Jeeps have a dashboard that is welded to the main body tub, so you have to lay on your back and do everything from underneath. It's a good idea to wrap a cloth around the steering column to prevent scratching the paint. Here, the passenger's end of the dashboard is supported by a cardboard box so that it does not bend.

With the bolt-on dashboard system, you can actually sit in the driver's seat where it's comfortable and work on the electrical wiring. Do not install the steering wheel until after the wiring and dashboard are complete. You may want to install the steering wheel loosely so you can move the Jeep around, but don't tighten the steering wheel nut. That way, you can remove it again while working on the dashboard.

With flat-fender Jeeps, you have to work from underneath the dash. It is recommended that you do not install the seats until after the wiring is done. This makes access easier. A favorite tool for doing electrical wiring under the dashboard is called a pillow. It comes in handy when lying on your back on the floor of the Jeep. Here, you also see a box of assorted electrical connectors. Get a variety and keep them on hand when doing the wiring.

To prevent having to lie on your back for long periods of time, preassemble the dash wiring because the later CJ speedometer mounts behind the dashboard. You can install most of the wiring to the speedometer before you go near the Jeep.

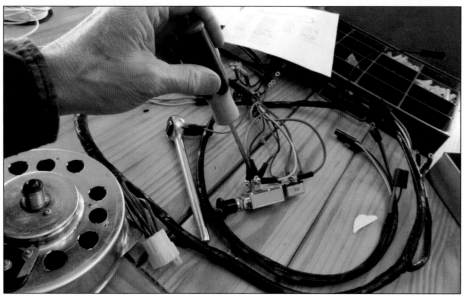

In addition to wiring the speedometer, you can wire the light switch and the ignition switch while still outside the Jeep. Both switches install from behind the dash.

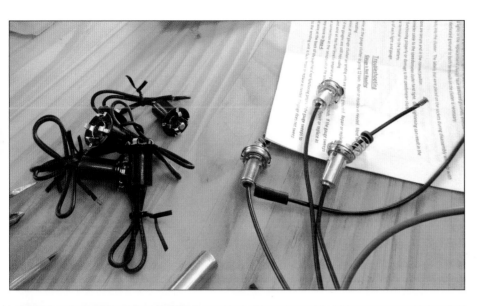

The speedometer and wiring harness are manufactured by two different companies. Sometimes the light bulb sockets that come with the wiring harness do not fit firmly in the back of the speedometer. You can order an extra set of sockets that fit tighter and splice them onto the appropriate-color wire found on the instructions sheet.

Neatness counts! The white-colored wire tie-downs mount with a screw or nut and bolt. This keeps the wiring harness in place. They are available in three different sizes.

I like to assemble the headlights and marker lights onto the grille before it is installed on the Jeep. The front headlight wiring harness is easier to install on the lights before the grille is mounted on the Jeep.

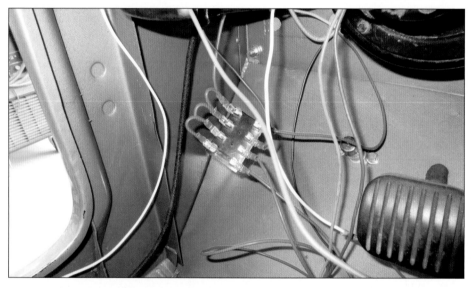

Many Jeeps have options and accessories that the wiring harness does not have provisions for. Therefore, an extra fuse box should be added to your system. Simply run a 10-gauge wire from the accessory terminal on the ignition switch. This way, power to the accessories will only be on while the ignition switch is on. Loop the positive wire to each terminal on the fuse box. The wires on the other side of the junction block go to your accessories, such as the heater, radio, electric wipers, etc. Use a 15- or 20-amp fuse depending on what the accessory instructions call for. You can also use the more modern flat-blade style of fuses. Both of these fuse styles and fuse boxes are available at local auto part stores. Mount the new fuse box on the firewall under the dash so it is out of the way but still accessible.

adding accessories, such as radios, lights, winches, etc. A new wiring harness eliminates many future headaches.

Each new wiring harness comes with a complete set of instructions and is assembled according to your Jeep's year and model. When you order the new harness, specifically tell the vendor the year, model, engine, alternator or generator, and electrical system (6 or 12 volt). Provide as much detail as possible. It is surprising how easy it is to install a new wiring harness once it is assembled for your specific Jeep. The wires will be labeled for each circuit.

From the early 1950s through the 1960s, 1970s, and 1980s, the wiring harness came in three main pieces: the main dashboard harness to the engine and ignition, the headlight harness, and the taillight harness. All three harnesses meet by the firewall to be connected to the gauges and switches.

The wiring harness for the MB and GPW comes in many small sections. When you follow the instructions, it may take a little longer to install, but it is not difficult to do.

The 1941–1945 Willys MB and Ford GPW models built during World War II had 6-volt electrical systems. The later military M38 and M38-A1 models had 24-volt electrical systems to meet NATO standards. Civilian Jeeps had 6-volt systems through 1956. Jeep CJs had 12-volt systems from 1957 onward.

The 6-volt systems cranked the engine over slowly, and the batteries died quickly. So, people began converting their Jeeps to a 12-volt system when the parts became available. Most people kept the 6-volt starter and just changed the generator, coil, voltage regulator, and battery

Laying out the wiring harness is a very technical process, so give yourself as much room as possible. Lay out the wiring, switches, gauges, and tools and then follow the instructions supplied by the manufacturer.

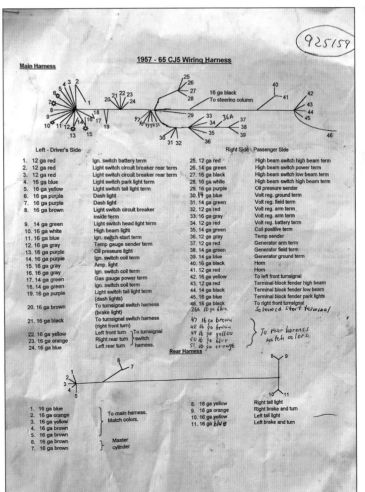

This diagram is a typical sample of what you receive with your wiring harness. It shows you each step in both numbered and color-coded instructions. Most people are afraid to tackle the electrical wiring in a Jeep restoration. With this illustrated schematic, the manufacturer makes it user friendly.

When installing the wiring harness for a Willys MB or Ford GPW, it is a little more technical than a civilian CJ. The newer civilian Jeeps have a three-piece wiring harness that simply plugs in when you follow the instructions. The MB and GPW wiring harness comes with 20 pieces. A valuable tool for installing this is a can of clothespins. You can attach the paper tag, part number, and description to each piece of wire. Then, follow the instructions and simply remove the clothespin once you attach the wire.

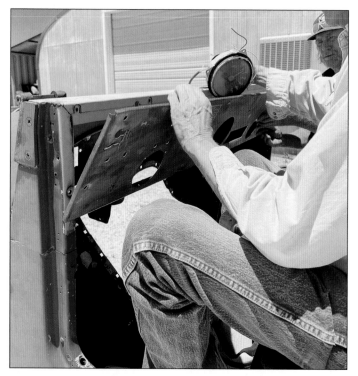

On 1941–1956 Jeeps, you'll find a five-gauge dashboard. The gauges mount from the front, but you can still pre-wire everything. As you install each gauge, simply remove the old wire and install the new wire to hook it back up. Do not install the speedometer until the very last. This makes it easier to stick your hand through the speedometer hole to reach the other four gauges.

to 12-volt parts. The 6-volt starter cranks the engine faster with a 12-volt electrical system. Some people get years out of a 6-volt starter, but don't be surprised if it burns out faster than a 12-volt starter.

Another common upgrade is changing from the generator to an alternator. Again, if you are planning on doing this, tell the manufacturer when ordering your unique wiring harness.

New, modern upgrades to Jeep electrical systems include the following:

- A one-wire alternator with a voltage regulator bolted inside. They are easy to install by just hooking up the one hot wire, and away you go.
- Low gear ratio starters. The starter spins faster and less cranking is needed. These new starters are very popular and make a world of difference.

Speedometer Repairs

The speedometer is a very simple mechanical system in your Jeep. If it does not work, it's likely one of three things:

- Most commonly, it's the speedometer cable that goes bad. The cable has threaded ferrules on both ends and is easy to replace.

With V-6 and V-8 engines, a separate wiring harness plugs into the dash harness. If you are doing a body-off restoration, it is easier to install this electrical harness first. There are many different types of alternators and generators. When you order the wiring harness, always explain to the manufacturer what equipment you are using in your Jeep. This way, you know that the wires will match up.

The 6-volt wiring harnesses used from 1941 to 1956 are cloth wrapped and color coded with dots.

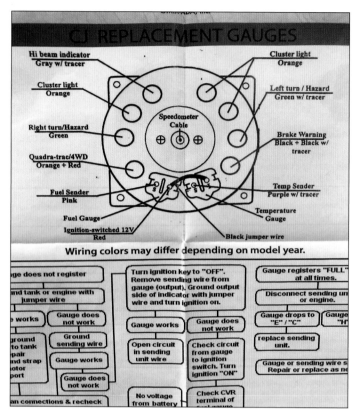

CJ REPLACEMENT GAUGES

Hi beam indicator
Gray w/ tracer

Cluster light
Orange

Cluster light
Orange

Left turn / Hazard
Green w/ tracer

Right turn/Hazard
Green

Speedometer
Cable

Quadra-trac/4WD
Orange + Red

Brake Warning
Black + Black w/ tracer

Fuel Sender
Pink

Temp Sender
Purple w/ tracer

Fuel Gauge

Temperature
Gauge

Ignition-switched 12V
Red

Black jumper wire

Wiring colors may differ depending on model year.

...ge does not register	Turn ignition key to "OFF". Remove sending wire from gauge (output). Ground output side of indicator with jumper wire and turn ignition on.	Gauge registers "FULL" at all times.
...nd tank or engine with jumper wire		Disconnect sending un... or engine.
...works	Gauge does not work	Gauge works	Gauge does not work	Gauge drops to "E" / "C"	Gauge ..."H"
...round to tank ...pair nd strap otor port	Ground sending wire	Open circuit in sending unit wire	Check circuit from gauge to ignition switch. Turn ignition "ON".	replace sending unit.
	Gauge works	Gauge does not work		Gauge or sending wire s... Repair or replace as ne...
...an connections & recheck	No voltage from battery	Check CVR terminal of...	

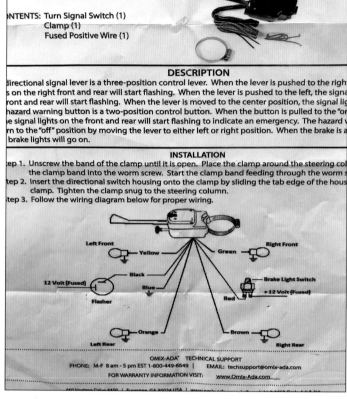

...NTENTS: Turn Signal Switch (1)
Clamp (1)
Fused Positive Wire (1)

DESCRIPTION

...directional signal lever is a three-position control lever. When the lever is pushed to the righ...
...s on the right front and rear will start flashing. When the lever is pushed to the left, the signa...
...ront and rear will start flashing. When the lever is moved to the center position, the signal lig...
...hazard warning button is a two-position control button. When the button is pulled to the "or...
...e signal lights on the front and rear will start flashing to indicate an emergency. The hazard ...
...rn to the "off" position by moving the lever to either left or right position. When the brake is a...
...brake lights will go on.

INSTALLATION

...ep 1. Unscrew the band of the clamp until it is open. Place the clamp around the steering col...
the clamp band into the worm screw. Start the clamp band feeding through the worm s...
...ep 2. Insert the directional switch housing onto the clamp by sliding the tab edge of the hous...
clamp. Tighten the clamp snug to the steering column.
...ep 3. Follow the wiring diagram below for proper wiring.

Left Front — Yellow   Green — Right Front

12 Volt (Fused)   Black

Flasher   Blue   Brake Light Switch

+12 Volt (Fused)

Red

Left Rear — Orange   Brown — Right Rear

OMIX-ADA® TECHNICAL SUPPORT
PHONE: M-F 8 am - 5 pm EST 1-800-449-6649 | EMAIL: techsupport@omix-ada.com
FOR WARRANTY INFORMATION VISIT: www.Omix-Ada.com

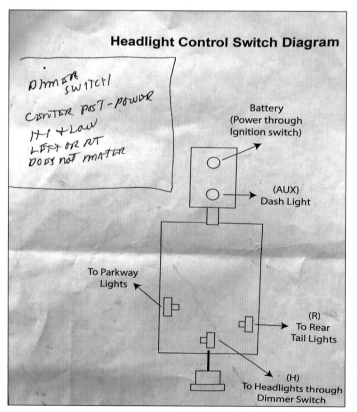

Headlight Control Switch Diagram

DIMMER SWITCH

CENTER POST-POWER

HI + LOW
LEFT OR RT
DOES NOT MATTER

Battery
(Power through
Ignition switch)

(AUX)
Dash Light

To Parkway
Lights

(R)
To Rear
Tail Lights

(H)
To Headlights through
Dimmer Switch

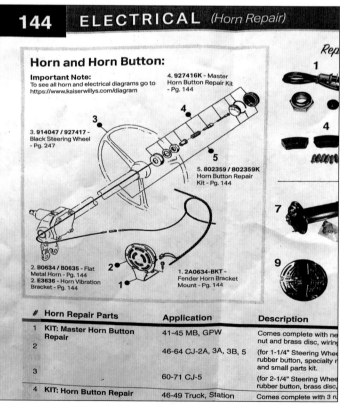

144 ELECTRICAL *(Horn Repair)*

Horn and Horn Button:

Important Note:
To see all horn and electrical diagrams go to
https://www.kaiserwillys.com/diagram

4. 927416K - Master
Horn Button Repair Kit
- Pg. 144

3. 914047 / 927417 -
Black Steering Wheel
- Pg. 247

5. 802359 / 802359K
Horn Button Repair
Kit - Pg. 144

2. B0634 / B0635 - Flat
Metal Horn - Pg. 144
2. E3636 - Horn Vibration
Bracket - Pg. 144

1. 2A0634-BKT -
Fender Horn Bracket
Mount - Pg. 144

#	Horn Repair Parts	Application	Description
1	KIT: Master Horn Button Repair	41-45 MB, GPW	Comes complete with ne... nut and brass disc, wiring
2		46-64 CJ-2A, 3A, 3B, 5	(for 1-1/4" Steering Whee... rubber button, specialty r... and small parts kit.
3		60-71 CJ-5	(for 2-1/4" Steering Whee... rubber button, brass disc,
4	KIT: Horn Button Repair	46-49 Truck, Station	Comes complete with 3 ru...

As you purchase each piece of equipment or accessory, it should come with its own installation instructions or schematics. Often, the supplier's catalogs have the schematics too, such as this one for the horn button.

- If it's the speedometer gauge itself, it's easy to change. In flat-fender Jeeps with individual gauges, two nuts and a clamp hold the speedometer gauge in the dashboard. There are no other wires or lights.

  The later round-fender CJs have a large single instrument cluster with several gauges in it. The large instrument is held in place from the back of the dashboard by four nuts and a few wires for the gas gauge and temperature gauge. All the dash light sockets simply unplug.

- The speedometer gear is located in the transfer case and can be removed and replaced with a wrench. If you change the tire size, you can replace the speedometer gear with a different ratio to compensate. This will make the speedometer accurate again.

Take the old speedometer cable out and compare the ends to the new one. The M38 and the M38-A1 cable look different than the civilian CJs. Make sure both cable ends are clean and that you line up the square cable end to the hole.

The speedometer cable can be removed from the transfer case with a wrench. Inspect the cable and make sure it turns and is not broken inside. Clean off the threads before reinstalling the cable.

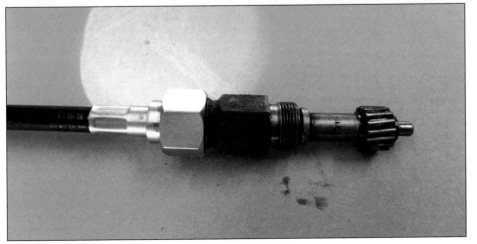

The speedometer gear can be removed from the speedometer cable. If you have changed the tire size on the Jeep, the speedometer gear can be replaced with a new one that provides a different ratio. A new gear with the correct ratio makes the speedometer accurate again.

TOPS

Military Jeeps had waxed canvas tops, seat covers, and accessories. The waxed canvas material has a unique smell that brings back memories for most veterans. The U.S. Army provided "summer" tops and "winter" tops depending on the theater in which the Jeeps operated. Summer tops had no side curtains on the rear quarters and no doors. Winter tops had side curtains and canvas doors.

Words were sometimes stenciled on the canvas tops to warn other drivers that these American vehicles were slow and left-hand drive.

A number of companies manufacture canvas reproduction tops and seat covers.

Civilian Soft Tops

Civilian Jeeps also came with a variety of soft top options. However, they were typically made with "leatherette vinyl." The CJ-2A had brackets welded to the left side of the body for the soft top's curved bows. The cross bows were attached to the windshield frame with clamps.

Later CJ models did not have storage for the soft top bows (frames). The bows were hinged and laid back along the edge of the body tub. The

soft top material could be left on or removed.

Steel rods were formed to produce a door frame and the soft top vinyl was stretched over the frames. These were called "skins." The soft tops and doors came with clear vinyl

windows that zipped open or closed. Reproductions are still available today.

If you have a Jeep without a top and decide to add one, be aware that these tops shrink over time. When placing the snaps on the body, attach

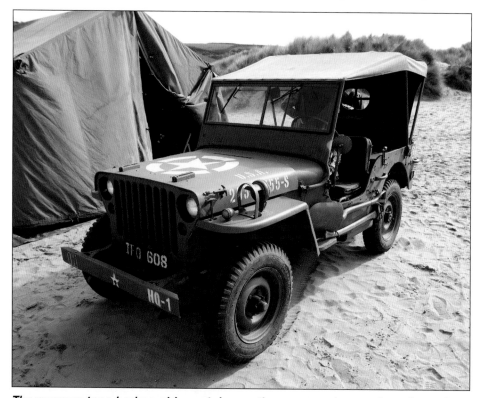

The summer tops had no side curtains on the rear quarters and no doors, but they were easier to get into and the visibility was a little better. The faster the GI drove, the better the breeze. (Photo Courtesy Rodney Hutchinson/Shutterstock.com)

The winter tops had side curtains and canvas doors. These were priceless for GIs driving in European winters without heat. (Photo Courtesy Howard Pimborough/Shutterstock.com)

Other drivers had to be warned about the American GI driving slowly, having the steering wheel on the "wrong" side, and having no turn signals. (Photo Courtesy Krzysztof Pazdalski/Shutterstock.com)

them higher on the body than the instructions indicate so that you will still be able to stretch the top in the future and attach the snaps.

The same process is needed for the zippers on the rear window. Try to leave a bit of slack for the zippers so that as the material shrinks you will be able to open and close the zippers. With time, weather, and grit, the zippers may become difficult to operate. Wash the top and zippers and then apply zipper lubricant or beeswax.

The clear vinyl windows eventually get scratches and become hazy. Vinyl polish is available to remove the haze. Check with the manufacturer of your top for the specific product.

Civilian Hard Tops and Hard Doors

Some early Jeep owners lived in cold regions and wanted better tops: ones with sturdy cabs and glass windows that could be defrosted or scraped. Several companies began to offer aluminum tops and doors with windows. Some were well built and kept the heater's warmth inside. Others were not very weather-tight. They were easily bolted on in the fall and removed in the spring. The better-built aluminum tops that were cared for and stored indoors have survived and can still be found today.

In 1976, with the introduction of the CJ-7, Jeep offered optional hard tops made of fiberglass and removable hard doors with roll-up glass windows, door handles, and locks.

If your CJ-7 or CJ-8 has a hard top and hard doors, you'll probably need to do a bit of maintenance as part of your project. While the hard top is off the Jeep, inspect the gasket that goes between the hard top and the body tub. They are easy to replace, and now is a good time to replace it if necessary. Also, look at the rubber gaskets around the side glass windows. They can dry rot and crack over time. The gas struts that prop the rear window open can weaken with age and may need to be replaced too. Last, consider whether the paint on the fiberglass top has faded and needs to be repainted.

If the hard doors on your Jeep CJ-7 or CJ-8 are original, they probably need some care as well. It's fairly common for the driver-side door to develop a crack in the steel door shell below the vent wing. The frequent closing of this door causes the door shell to flex and eventually crack. Welding the crack and grinding it smooth may give it more years of service. Paint the door shell when the body tub is painted to save cost.

The two door-hinge brackets on the body have bushings that allow the door to swing open and close smoothly. However, they wear out, allowing the door to sag and not latch properly. Brass replacement bushings are readily available from online Jeep parts suppliers.

Door handles and latches also wear and can become misaligned. Replacements are available and restore proper operation to the door. The same is true for the lock linkage and interior door handles. Lock cylinders on the outside of the door can become loose and are easily replaced once you remove the interior door panel.

Interior door panels can deteriorate due to exposure to the weather. Replacements are available in the standard Jeep colors. Window regulators raise and lower the glass windows. If they don't function properly, replacements are available.

Willys began building the first civilian Jeeps in late 1945 and named them the CJ-2A. Among the changes made for civilian use was a switch from a heavy, waxed canvas top to a vinyl top. The vinyl top was lighter, more flexible, and easier to take down or put up. Willys even added pockets and brackets to store the bows.

The aluminum fabrication industry thrived during World War II as it supplied aircraft manufacturers. After the war, small companies saw a demand for an aluminum Jeep top, and each made their own version. Some were simple, while others were pretty well engineered. Some of these tops can still be found today. (Photo Courtesy Seth Hensley)

Finally, the weatherstripping around the perimeter of the door, the glass wipes on each side of the glass window, and the glass edge channel in the door frame may need attention. Replacing them reduces the whistling from the wind and keeps the weather outside and the heat or air conditioning on the inside.

From 1946 through 1975, Jeep had only offered soft tops made of vinyl. This is part of what made a Jeep CJ iconic. Everyone recognizes a Jeep from its grille and soft top. However, when they added the optional fiberglass hard top, they made the Jeep much more practical and comfortable. Jeeps were driven more often in cold climates with this top.

Having hard doors was a major step forward for the Jeep CJ models. Every Wrangler since has had a hard door option. They are great for keeping passengers warm and dry in cold, wet, or muddy conditions. However, in time, they require maintenance. Window regulators fail, door hinge bushings wear, and gaskets crack. All of these are relatively easy to fix, and the parts are readily available.

Jeep doors have pins that drop into the hinges. Over time, the brass hinge bushings wear out and need to be replaced. Worn bushings can be drilled out and new ones tapped into place.

Coker Tire
2101 Roseville Ave.
Chattanooga, TN 37408
1-800-959-1900
cokertire.com

Collins Bros. Jeep
3101 W. FM 544
Wylie, TX 75098
972-442-6189
cbjeep.com
Specialty: Jeeps 1976–present

J&W Auto Wreckers
8628 Antelope North Rd.
Antelope, CA 95843
1-800-924-9732
jwjeep.com
Specialty: Jeeps 1941–present. The
company's yard has more than 800
Jeeps.

Kaiser Willys Auto Supply
114 Bolton Ct.
Aiken, SC 29803
1-888-648-4923
kaiserwillys.com
Specialty: Jeep parts 1941–1971

Midwest Willys
3706 S. 100 E.
Crawfordsville, IN 47933
765-362-2249
Email: midwestweillysllc@gmail.com
Specialty: 1941–1967 Jeeps (mostly
used parts)

Peter Debella Jeep Parts
1319 Pulaski St.
Riverhead, NY 11901
631-874-8660
debellajepparts.com
Specialty: military Jeeps 1941–1970

Portrayal Press LLC
P.O. Box 944
Branchville, NJ 07826
855-339-0382
Email: sales@portrayal.com
Specialty: Service manuals for all Jeeps

Walk's 4WD
700 Cedar St.
Bowmanstown, PA 18030
610-852-3110
walks42d.com
Specialty: Jeeps 1941–1980